Pride and Vision

The Canadian Institute of Mining, Metallurgy and Petroleum, 1898–1998

E. Tina Crossfield

Peter Tarassoff
Chairman, CIM History Editorial Board

Sponsored by the CIM Centennial Corporation

Published by the Canadian Institute of Mining, Metallurgy and Petroleum
Montreal, Quebec, 1998

Dépôt légal, Bibliothèque nationale de Québec and the National Library of Canada, 1998.

Canadian Cataloguing In Publication Data

Main Entry under title:
Pride and Vision. The Canadian Institute of Mining, Metallurgy and Petroleum, 1898-1998.

Crossfield, E. Tina, 1955-
Includes bibliographical references.
ISBN 0-919086-98-5

1. History - Mining, metallurgy and petroleum - Canada. 2. History of Canadian science and technology, 3. Biography

Canadian Institute of Mining, Metallurgy and Petroleum
Institut Canadien des mines, de la métallurgie et du pétrole
3400, boul. de Maisonneuve O., bureau 1210
Montréal, Québec, Canada H3Z 3B8

Printed in Canada

Dedication

To the Men and Women of the
Canadian Institute of Mining, Metallurgy and Petroleum
who helped create a national organization
of great distinction.

Sponsors

The Canadian Institute of Mining, Metallurgy and Petroleum gratefully acknowledges the generous support of the Honorary Members of the CIM Centennial Corporation.

Alcan Aluminium Limited
ASEA Brown Boveri Inc.
Atlas Copco Construction and Mining
Bank of Montreal
Barrick Gold Corporation
Cambior Inc.
Caterpillar and its Canadian Dealers
CIBC / Wood Gundy
CN (Canadian National)
Cominco Ltd.
Compagnie Minière Québec Cartier
Cummins Engine Company Inc.
Dyno Nobel Ltd.
Falconbridge Limited
Hitachi Construction Machinery
Hydro-Québec
ICI Explosives Canada
INCO Limited
Iron Ore Company of Canada
Komatsu Mining Systems Inc.
Ministère des Ressources naturelles du Québec
Natural Resources Canada
Noranda Inc.
P&H Mine Pro Services
Placer Dome North America
Royal Bank of Canada
Scotiabank
Secoroc Inc.
Syncrude Canada Ltd.
Tamrock Canada Inc.
Teck Corporation
The Toronto-Dominion Bank

Honourable Jean Chrétien, Prime Minister of Canada, Honorary Patron
René Dufour, Chairman, CIM Centennial Corporation
Giorgio Massobrio, Chairman, CIM/CMMI/MIGA Montréal '98

CONTENTS

PREFACE

The challenge of embarking on a project of this scope is to define the limits of investigation. There is always additional research which can be done, and other people who could, and perhaps, should have been interviewed. Time is generally the most limiting factor. The enormity of information on some topics, and the scant amount on others, indicates that there is still much history waiting to be written on Canadian mining, metallurgy, and petroleum.

The primary sources for this book were the Institute's publications, spanning 100 years, and Institute membership files and minutes of Council meetings, many of them located at the National Archives of Canada in Ottawa. Other important sources were the journals of the Institute's predecessor organizations, and mining periodicals such as the *Canadian Mining Review*, *Canadian Mining Journal*, and *Northern Miner*. The accounts of international professional societies have also been valuable.

One of the best ways to understand the history of an organization is through interviews. I wish to thank the following people for allowing me that opportunity: William Almdal, E.H. Caldwell, Noel Cleland, Jean-Paul Drolet, René Dufour, John E. Dutrizac, George Furnival, William Gilchrist, George Govier, Aubrey Kerr, John Laffin, Lindsay Milne, the late Alfred Lloyd Penhale, Walter Riva, C. Bruce Ross, Bill Stanley, Peter Tarassoff, John Walsh, and Don Worth. I also spoke with Yvan Jacques, Gordon Tapp, and Gordon Skilling. Several CIM staff members were extremely helpful: Perla Gantz and Serge Major at the Secretariat in Montreal; and Catherine Buchanan in Calgary. I am particularly grateful to Leon Monpetit for assisting with research on the Branch histories.

Much of the credit for the realization of this project is due to the efforts of Peter Tarassoff, Chairman of the Editorial Board, a skillful editor and a patient consultant on many topics. The members of the Editorial Board, Jean-Paul Drolet, John Dutrizac, Faithi Habashi, John Jambor and Aubrey Kerr, reviewed early drafts of the manuscript. The staff at the National Archives of Canada accommodated every request.

R.J.C. MacDonald prepared a history of the Canadian Mineral Processors. Richard Jauron and Roger Laliberté helped locate rare documents relating to the early asbestos industry in Quebec, and Cyrille Dufresne loaned a valuable album of clippings and photographs assembled by his father, A.O. Dufresne. David Lockwood provided invaluable technical assistance.

This book is a Centennial project funded by the CIM Centennial Corporation, to whom I am grateful for the opportunity.

I would like to thank the following people for their continuous support: Marianne Ainley, Gerry Boos, David Finch, Anne-Marie Weidler-Kubanek, George Kubanek, and Andrea Winlo. And my heartfelt thanks go to Kristin and Stefan Kunze, and to my husband, Harald Kunze, for their unfaltering belief.

INTRODUCTION

Many Canadians equate settlement and frontier in the late nineteenth century with the advance of the railways. In reality, it was a combination of mining and the railways which laid the foundation for numerous communities across the country. Mining camps commonly predated other measures of civilization, and were it not for the demand for valuable minerals, road and rail might not have deviated beyond the more direct routes of the surveyors. Canada was then an open land of fewer than 5.3 million people. Most men worked in agriculture, mining, logging, and fishing, while manufacturing supplied tools and other necessities. The majority of women devoted their days to homemaking, teaching, and health care, and attempted to weave together the social fabric so necessary for towns and cities.

Curiously, there are also misconceptions about science and technology during the same period. Most people attribute the generation of new knowledge to educational and government institutions. Although this is true to some extent, it does not provide a complete picture. Those who participated in natural history by keeping meticulous records and publishing their findings, along with those who were actively engaged in industries such as mining, also made valuable contributions. Early science in Canada was based on economic value, and governments considered the land as a warehouse of resources where an understanding of soils, minerals, and farming practices would turn a profit. Yet government funding for scientific research was limited, and if results were not immediately useful, then the more far-sighted projects were abandoned. As few paid scientists had the time or institutional support to carry out investigations beyond the scope of their duties, they often relied on information from external sources. One such source was the mining and metallurgical industry whose need for new or improved processes required much innovation. A common practice for university professors at the time was to act as technical consultants for industry, and carry out laboratory tests on its behalf. The constant efforts of people engaged in both industry and academia helped Canada to usher in the modern age.

The development of regional mining organizations, leading to the establishment of the present-day Institute, took place over a surprisingly short period. Members of the General Mining Association of the Province of Quebec, in 1891, and the Mining Society of Nova Scotia, in 1892, had both gathered to discuss problems within their industries and to share technical and scientific information through publications and meetings. Within ten years, these two associations, along with the Ontario Mining Institute (1894), had joined together as the Federated Canadian Mining Institute (1897), one of whose missions was to confront the government on legislative issues. In 1898, the reorganized Canadian Mining Institute (CIM) claimed membership in Nova Scotia, Quebec, Ontario, and British Columbia, "making a practically unbroken union from Atlantic to Pacific." Later, CIM deferred its lobbying activities to other groups such as the Mining Association of Canada, preferring to focus on the transfer of knowledge and experience, and the fostering of good relations with other professional mining groups worldwide. In addition, the Institute was often asked to advise the government on technical matters, and it played a key role in the establishment of the federal Department of Mines (1907).

The American Institute of Mining Engineers (AIME, 1871) held some of its annual meetings in Canada before any similar groups existed in this country, and its friendly exchange had a direct influence on the formation of the CIM. In return, Canadians have always been active in AIME, occasionally serving as officers. This connection has its own historical roots. From the mid-nineteenth century, there was a noticeable shift in the transfer of science and technology away from Britain with which Canada had initially shared information and ideas. Even though Canada was foremost a British colony which reflected Victorian ideals, traditions, and administrative systems, people increasingly looked to the United States, and to institutions such as AIME for affiliation and scientific exchange. In later years, there was sometimes competition between CIM and AIME for members, but the conscientious efforts of both Institutions have ultimately prevailed.

Although the history of the Canadian Institute of Mining, Metallurgy and Petroleum parallels the growth of the mineral industry, the progress of science and technology, and the advance of the railways, this book is mainly about people. Of the Institute's original 23 Founders, some came to Canada from Britain, France, Scandinavia, and the United States. Most were well-educated, listing their occupations as mine owners and operators, engineers, professors, prospectors, and civil servants. In 1898, anyone was welcome to join the Institute regardless of title or citizenship, as long as they were "engaged in the direction and operation of mines and metallurgical works, mining engineers, geologists, metallurgists or chemists, or any other person whom the Council saw fit to elect." In the 1950s, women scientists and engineers were officially welcomed into the membership ranks.

This book marks the first attempt to document and explain the development of the CIM. Although some articles on the history of CIM have appeared in the *Bulletin*, they have mainly relied on anecdotes and extracts from Institute publications. The best article to date remains the one written in 1910 by Secretary Mortimer-Lamb, who felt strongly that the "present membership is uninformed concerning the early history of the society, the causes that led to and resulted in its organization, or of the useful services it has rendered the mining industry since its inception."

One of life's certainties is that change is inevitable, and in order for an organization to remain relevant to its membership, it must be ready to adapt. In the earlier years, initiating change was sometimes difficult, and debate would continue over a period of years. Often it was a question of timing, and some Institute officers were more successful at altering the status quo than others. *Council minutes* record sound ideas that failed, and resolutions, unanimously passed, which were never implemented. It is not surprising that some of the problems faced by the Institute's Founders still present a challenge. These include finances, publishing, education, and striving to keep a large and diverse membership together.

This history is organized into four chapters. The main chapter is a chronological narrative of CIM's history from 1898 to 1998 and is divided into three sections. *The Early Years* (1898-1919) retraces critical events in the 1890s, and explains the factors leading up to the formation of the Canadian Mining Institute. Here, we discover B.T.A. Bell and the *Canadian Mining Review*, CIM's founders, how branches were conceived, and what challenges were facing the Institute before and after World War I. *The Middle Ground* (1920-1966) takes into account the Great Depression, World War II, a proposed Headquarters relocation, the launching of technical Divisions and Societies, and the first women members. In *The Modern Era* (1967-1998), some of the major events, decisions, and changes within CIM are highlighted, many of which will be familiar to current members. These include new publications, the cyclical patterns of finance and membership, the computerization of the national office, the environmental challenge, the reorganization of Council, and the Institute's identity.

Chapter Two is a short chronicle of the Branches, and their relationship to the Institute. Although historical information on most Branches remains sketchy, the accounts of a few can provide an insight into the rest.

From the first three Branches, formed in 1902, to the more than fifty operating in 1997, Branches continue to provide CIM with its grass-roots foundation. A capsule history of all known Branches is included in a chart format for easy reference.

Chapter Three highlights the history of the Institute's technical groups, beginning with the Industrial Minerals Division in 1940. It was followed by the Coal Division (1944), the Metallurgy Division (1945), the Geology Division (1945), the Metal Mining Division (1946), the Petroleum and Natural Gas Division (1949), the Mechanical-Engineering Division (1961), the Canadian Mineral Processors Division (1981), and the Mineral Economics Society of CIM (1992). Four of the older Divisions underwent name changes in succeeding years.

The fourth chapter is a brief look at the aura of the Annual General Meeting, which has always been the highlight of each Institute year, and an Epilogue looks ahead into CIM's future. Finally, two appendices provide short biographies of the Institute's Founders, Presidents, and Executive Directors.

In sifting through CIM's history, the question of membership resurfaces time and again. What does the Institute actually mean to members, how can it best meet their needs, and how can members contribute to its on-going success? Men and women working in industry have always taken pleasure in the sharing of knowledge and experience. Belonging to CIM means being part of an organization that is both welcoming and protective, one that recognizes achievements and awards excellence. It is like walking into a remote northern mining town and meeting an old friend from a different part of Canada. A handshake is either followed by laughter at the recounting of a memory, perhaps from the last Annual General Meeting, or a compassionate word upon hearing of a personal loss.

The Institute is about people, and membership in CIM has always been held in high esteem by those willing to accept a challenge. As the new millennium approaches, the organization of more than 12,000 members will need to adapt to new challenges. The camaraderie in which the Institute was first formed has held it together through war, economic depression, and rapid technological change. As in the past, the Institute's continued success will only be realized if men and women take an active interest in its well-being. Just as "Members are the Industry," the CIM is, indeed, the membership. It radiates pride in its past as warmly as it shines with a never-ending vision of its future.

CHAPTER ONE

Formation and Evolution of the Canadian Institute of Mining, Metallurgy and Petroleum, 1898 to 1998

Since its inaugural meeting in Montreal on March 4, 1898, the Canadian Institute of Mining, Metallurgy and Petroleum was destined to become one of the most important and influential professional organizations in Canada. Its historical roots can be found in the legacy of several far-sighted and diverse groups that were guided by innovative and experienced men. These include the Gold Miners Club of Nova Scotia, the Mining Society of Nova Scotia, the *Canadian Mining Review*, the General Mining Association of the Province of Quebec, the Ontario Mining Institute, the British Columbia Association of Mining Engineers, the American Institute of Mining Engineers, and the Federated Canadian Mining Institute. Each had a role to play in the two decades before the Dominion Charter of June 18, 1898 formally proclaimed the birth of the Canadian Mining Institute.[1]

[1]The Institute has existed under different names: the Canadian Mining Institute (1898-1920), the Canadian Institute of Mining and Metallurgy (1920-1990), and the Canadian Institute of Mining, Metallurgy and Petroleum (1990 to the present). Throughout the text, CIM is used as the common abbreviation.

THE EARLY YEARS: 1898-1919

The Founders

In the Act of Parliament of 1898, twenty-three men are named as founders of the Canadian Mining Institute (CIM). These and others individuals piloted the organization throughout the early years, when convincing the public that the mineral industry was of great importance, and that mining men were respectable citizens, was a difficult task.[2] Indeed, the founding members exemplified just how high mining men could aspire. One such individual was John E. Hardman, who assumed the central role as CIM's first president. An American by birth, he was a well-trained mining engineer from Boston's Institute of Technology, who went to Nova Scotia in 1883 to work in the gold mines. Ten years later, he helped to establish the Mining Society of Nova Scotia, serving as its first president in 1894. In great demand as a speaker, he lectured the mining students at McGill University in 1895 before opening his own practice as a mining consultant. The fledgling Institute could have had no greater leadership than that of Hardman.

The Institute had four vice-presidents who served concurrently. The first was George M. Dawson, Director of the Geological Survey of Canada (GSC) from 1895 until his untimely death in 1901. His father was Sir William Dawson, Principal of McGill University. At times accompanied by Richard George McConnell, another Institute Founder, George Dawson had explored and mapped the Yukon Territory and various parts of northern British Columbia for nearly a decade before the Klondike gold discovery of 1896. The 2nd Vice President was William A. Carlyle, a McGill graduate who was British Columbia's first Provincial Mineralogist. In 1898, he was the mine manager at Rossland for the British American Mining Corporation. John Blue, 3rd Vice President, and past president of the General Mining Association of the Province of Quebec (GMAPQ) in 1895, managed the Eustis Mining Company, a copper producer in Capleton, Quebec. Charles Fergie, fourth in the roster, had extensive experience in Nova Scotia's coal industry. A.W. Stevenson, a charted accountant at the Bank of Toronto Chambers in Montreal was treasurer.

[2] When George R. Smith briefly left his seat in Quebec's Legislative Assembly, he was criticized by the opposition "For leaving the House for two days to attend an affair of no importance." The "affair" was his election as Institute President in 1906. Smith was an M.L.A. from 1897 to 1908. *CIM Bulletin*, no. 731 (March 1973): 22.

In addition, the mining interests of four provinces were represented on the first Council by the following people, some of whom had experience in their own provincial associations. From Quebec, George A. Drummond of the Canada Iron Furnace Company became an Institute vice president the following year. Joseph Obalski, born in France, was a graduate of l'École des Mines and had been Quebec's Mine Inspector since 1881. John J. Penhale, a leading member of the General Mining Association of the Province of Quebec, and George R. Smith, were both involved in asbestos mining. The former was affiliated with the United Asbestos Company, and the latter was with Bell's Asbestos Company. Penhale and his father, Matthew, had designed and built the world's first asbestos mill at Black Lake. Smith's talents would lead him into politics, and his knack for diplomacy would serve the Institute well in the troubled years ahead. Albert P. Low, an energetic explorer with a consuming interest in natural history, was a McGill graduate and had worked for the Geological Survey of Canada since 1882; he would become its director in 1906.

The British Columbia representatives were John B. Hobson of the Cariboo Hydraulic Mining Company, William Blakemore of the Crow's Nest Coal Company, and Frank C. Loring, whose pioneering spirit was to lead him from the gold and copper mines of Rossland, to Cobalt and then to Kirkland Lake. Richard G. McConnell, the Provincial Mineralogist, was described as "a strong, rugged, and intrepid explorer, robust and of sturdy Scotch ancestry [and] a great reader of good books…"[3]

Representing Nova Scotia, Robert G. Leckie had immigrated to Montreal from Scotland in 1859, and first worked in shipbuilding before acquiring a keen interest in mining. By 1898, he had gained considerable expertise in copper, nickel, gold, and coal mining, and was about to become the resident consulting engineer for the Canadian Copper Company in Sudbury. When the Federated Institute was formed in 1897, Leckie was elected president, and in later years, he would affectionately be known as "Dean" of the Institute. The other Nova Scotia representatives were Wilbur L. Libbey, a broad-shouldered American who had arrived in Nova Scotia in 1894 to work for the Brookfield Mining Company, and Clarence H. Dimock, a dapper gentleman who operated the Wentworth Gypsum Company at Windsor. Henry S. Poole of the Acadia Coal Company in Stellarton was a leading figure in the Mining Society of Nova Scotia.

[3] Frank Dawson Adams, "The Late R.G. McConnell," *CIM Bulletin*, no. 361 (May 1942): 166.

Ontario's mining interests were represented by James McArthur of the Canadian Copper Company in Sudbury, Frederick T. Snyder of the Ottawa Gold Milling and Mining Company in Keewatin, and Archibald Blue, first Director of the Ontario Bureau of Mines, established in 1891. The fourth representative was Wm. Hamilton Merritt of Toronto, an Associate of the Royal School of Mines in London, England, who lectured in mining engineering at Kingston's School of Mining from 1893 to 1903. All of these men were recognized as Charter Members of the Institute.

Legacy of B.T.A. Bell

Historians have noted that in Canada, as well as in the United States and Britain, "the initiative for forming an association of professional mining men came from the media."[4] In this respect, the prime mover in Canada was Benjamin Taylor A. Bell, editor of the *Canadian Mining Review* since 1886.[5] The *Review* was an Ottawa publication, launched in 1879 by W. A. Allen and Grant Powell. Allen had some experience in the Ottawa district phosphate mines, and Powell was Canada's Under-Secretary of State for External Affairs. Both had a good sense of where the Canadian mining industry was headed. In 1883, the *Review* reported that:

> In Nova Scotia the gold mines are, in many instances, yielding more per man than those of any other gold region in the world... Some copper and iron mines are also paying well, though the latter mineral has not yet received attention commensurate with its presence... [but] her coal fields have, for many years, been a source of great wealth.[6]

Two years later, the *Review* observed that the Spring Hill Collieries were raising an average of 1,500 tons of coal daily.

So important were the coal fields of Nova Scotia that in 1885 the American Institute of Mining Engineers (AIME) held its annual meeting in Halifax. Formed in 1871, this organization promoted the development of

---

[4] A. J. Wilson, *The Professionals. The Institution of Mining and Metallurgy 1892-1992* (London, England: IMM, 1992). A.J. Wilson, *The Pick and the Pen* (London, England: Mining Journal Books, 1979).

[5] The *Canadian Mining Review*, hereafter the *Review*, changed its name to the *Canadian Mining Journal* in 1907.

[6] W. Schabas, "B.T.A. Bell, The Canadian Mining Review, and the Early History of Nova Scotia's Mining Industry," unpublished paper, 1977, 13 p.

new methods for assessing mineral deposits, and extracting and processing them efficiently. The AIME was keen to learn about mining engineering in Canada. The meeting was attended by 150 people and was held in the Nova Scotia Legislative Council Chamber. Several of the technical papers were authored by Canadians. Robert G. Leckie spoke on "The Coal Fields of Cumberland County," and E. Gilpin, Inspector of Mines for Nova Scotia, presented a paper on "The Nova Scotia Gold Mines."

The AIME gathering in Halifax may have been the initial spark that fuelled the formation of the Gold Miners Club of Nova Scotia two years later.[7] The first professional mining association in Canada, it was a direct ancestor of both the Mining Society of Nova Scotia (MSNS) and the Federated Canadian Mining Institute. The Gold Miners Club met in the office of the *Halifax Critic* on November 1887 to draft a constitution and by-laws. Elections for executive positions were held in February, and Benjamin Taylor A. Bell was named as the only honourary member. When the Gold Miners Club was reorganized as the Gold Miners Association in 1889, the *Review* was adopted as its official publication.

Until the middle of the nineteenth century, there was little commercial development of Canada's mineral resources. Coal mining had begun on Cape Breton Island in the 1720s, and the first officially recorded export of minerals from Canada was a shipment of coal to Boston in 1724. The mining and smelting of iron ore, and the manufacture of iron products, met the immediate needs of Canada's largely agrarian population. The first iron works was constructed at Quebec's Les Forges du Saint-Maurice in 1729, and other iron works sprung up in various parts of the province, as well as in Ontario and Nova Scotia. The first gold production came from placer deposits on Quebec's Chaudière River in 1823, and the first copper production was at Bruce Mines in Ontario from 1848 to 1850.

By the second half of the nineteenth century, the pace of mineral development rapidly accelerated. In the 1850s, the discoveries of coal at Nanaimo, and gold on the Fraser River and in the Cariboo district, led to widespread mineral development in British Columbia. By the 1890s, the production of silver, lead, and copper was booming in the mining districts of West Kootenay, Slocan, Nelson, and Ainsworth, and Rossland had become known for its gold and copper mines. On the east coast, Nova Scotia saw the first hard-rock gold mines in Canada when gold was found in Mooseland, Tangier, Sherbrooke, and other districts in the 1850s. In Quebec's Eastern

[7] Schabas, ibid.

Townships, the American Civil War stimulated a copper mining boom during the 1860s, and the mines in the Thetford Mines area began producing *pierre à coton* in 1878, and were destined to become the world's largest supplier of asbestos. Farther west in Quebec, the Lièvre River region was the centre of a phosphate (apatite) industry between 1875 and 1890.

In Ontario, Oil Springs ushered in the development of Canada's petroleum industry as early as 1862. The first hints of the vast mineral wealth of the Canadian Shield came when gold was found near Madoc in Hastings County in the late 1860s, followed by silver near Port Arthur and at Silver Islet on Lake Superior's north shore. The most significant development, however, was the discovery in 1882 of the great copper-nickel deposits in the Sudbury Basin. On the eve of the twentieth century, Canada was poised to become one of the world's great mineral producers and many of these developments were chronicled in the pages of the *Review*.

When Bell was appointed by the Dominion government in 1890 to accompany Alfred Selwyn, Director of the GSC, on a tour of the mining and industrial centres of Canada, he was already familiar with the mining scene. Having accepted the editorial chair of the *Review* four years earlier, after owner Grant Powell had gone to England, Bell had not only learned about the problems of the industry first hand, but had become its strongest advocate and spokesman. Through his intimate association with key individuals in the industry, he spearheaded the organization of four provincial mining associations and helped bring about their federation in 1896, which emerged two years later as the Canadian Mining Institute. Bell was especially concerned with misrepresentation and dishonesty by mining promoters and specualtors, and by what he perceived as unfair government tactics. His feisty editorials in the *Review* earned him at least one threatened libel suit per year, although there is no evidence that any proceeded to the courts.

Bell was born in Edinburgh, Scotland, in 1863, the son of Dr. David Bell and Margaret Taylor from whom he received his compound name. While a student at Stewart's College, he began using the initials B.T.A. in order to distinguish himself from another lad with an identical name. As a young man, he was convinced that *seeing life* was better than *schooling*, and after working for a short time for an engineering firm, he caught an emigrant ship to Halifax in 1882. He attempted to settle on the prairies near Brandon, Manitoba, but soon discovered that farming was not his forte, and he found interim work as a sign painter. In 1883 he was hired as a freight clerk by the Canadian Pacific Railway in Winnipeg. The following year, he enlisted in the Manitoba Contingent of the Canadian Voyageurs and served in the Sudan, receiving the Egyptian Medal and Khedive's Star for his service. In 1885, he returned to Canada where he took an editing job with the *Athletic News*, and wrote articles for the Montreal *Gazette* and Montreal *Witness*. Bell was an outstanding cricketer and while in Ottawa for a cricket match, he was offered an editorial position with the *Canadian Mining Review*. Bell's name first appeared on the journal's masthead in 1887, the very same year the Gold Miners Association was formed. A short time later, he was a joint owner with Powell. By 1896, the journal was his.[8]

Although most pages in the *Review* were devoted to correspondence from across the country, and to technical papers gleaned from a variety of sources, Bell used the editorial pages to comment on issues that caught his attention. One time he took it upon himself to criticize inaccurate information about the mining industry appearing in a rival paper called the *Halifax Critic*, and published their inconsistent reports alongside his own opinions. Bell and C.F. Fraser, the *Critic*'s editor, had already been bitter competitors for several years. Fraser, who had earlier tried in vain to organize the mining engineers, lost out to Bell when the *Review* became the Gold Miners Association official organ. The takeover was never forgiven.[9]

The principles of honesty and integrity within the mining sector, along with the public right to know about risky or fraudulent investments, became one of the foundations for what was to become the CIM. The best medium for dispensing information of this nature was through the press, and there is no doubt that Bell had the best skills as a writer to deliver it. However, Canadian mining engineers had a much larger goal in mind: sharing their knowledge and experience not only among themselves, but with their colleagues elsewhere, particularly those associated with the already prestigious AIME south of the border. The AIME meeting in Halifax in 1885 had made a favourable impression on the Canadians, and had provided them with a working model. For the next two decades, whenever the Institute was unsure of how to proceed, its members would refer to the AIME by-laws for guidance. But the single major factor which accelerated their cause was events taking place in government.

[8] W. Schabas, "B.T.A. Bell - Canadian Mining's Spokesman," Centennial Series, *Canadian Mining Journal*, vol. 99, no. 6 (June 1978): 23-30.

[9] For a longer description of Bell's interaction with Fraser, see W. Schabas, "B.T.A. Bell, the Canadian Mining Review and the Early History of Nova Scotia's Mining Industry", ibid.

General Mining Association of the Province of Quebec

The foundation of the future Institute was laid on January 23, 1891, when a meeting of Quebec mine owners and operators was urgently called by B.T.A. Bell to devise a plan to combat legislation considered to be unjust and therefore threatening to the mining industry. Before the assembly was adjourned, it had adopted a resolution that formed the basis of the General Mining Association of the Province of Quebec (GMAPQ), under the capable guidance of the Hon. George Irvine, President of Johnson's Asbestos Co. There was some discussion among the fifteen or so delegates present that afternoon, in a small room at the Windsor Hotel in Montreal, on whether or not their interests were similar. Some believed, for example, that the government's method of issuing land patents was different for those mining asbestos than for other minerals, in which case the phosphate mine owners should organize by themselves. Irvine was quick to point out that in their current state of disunity, little would be accomplished, and if they formed smaller groups, their efforts to challenge the government would be futile. The only real solution lay in a "well-organized, thoroughly representative, and well equipped association."[10] After an exhaustive discussion of the implications, a motion to join forces was carried without further dissent. A committee was nominated to write a constitution, and the election of officers took place on January 30, 1891. Irvine was elected President, and Bell, Secretary.

For many struggling mine operators, it was a formidable challenge to manage a productive mine, stay ahead of the competition, and deal with onerous legislation, such as the new mining law recently introduced by Quebec's Mercier government, which not only levied a three percent royalty on mineral production, but was also tantamount to a confiscation of mineral rights. Until 1891, no general forum had existed to discuss problems inherent in the industry, and often the success of one operator was viewed by another with envy rather than praise. The sharing of technical information was rare, and few miners understood the benefits that education and public awareness could have on their own enterprise. A common cause was desperately needed to pull these men together, and this opportunity gelled in the form of the "ill-considered, impractical, and piratical attack upon [their] vested interests."[11] The GMAPQ countered

[10] "Special meeting, Montreal, 23rd January 1891," *Journal of the General Mining Association of the Province of Quebec* [*JGMAPQ*], vol. I (1891-92-93): 1-2.
[11] Robert C. Adams, "Preface," *JGMAPQ*, vol. I (1891-2-3): iii-iv.

the Mercier law, as it came to be known, by urging its members to refuse to complete forms and quarterly reports requested by the government until their petition to have the law disallowed had been addressed by the Governor-General-in-Council.[12]

Their collective outrage stemmed from issues of ownership as well as the royalty on mineral production. Under legislation passed in 1874, buyers of Crown land automatically gained the rights to any minerals, except for gold and silver. In the amended statutes of 1880, if minerals were discovered on the land, the owners could protect their interests by paying the difference between its value as a mining property and what they had originally paid. The problem with the Mercier law was that all mineral rights, whether on private or public lands, reverted to the Crown. Anyone could obtain an exploration permit, and if they discovered a mineral deposit, could make an offer to the Crown to purchase the mining property. If the land was privately held, the owner could either match the offer, or forfeit the land. The outrage of the mine owners was that land and mineral rights which they had previously purchased in good faith, could now be effectively confiscated.

B.T.A. Bell, in his infinite good judgement, had already sent a copy of the new Quebec mining law to Rossiter W. Raymond, Secretary of AIME. His reply was one of shock:

> I did not deem it possible that the legislature of any civilized country could at this day be induced to enact a measure so barbaric in its injustice and unwisdom... the new law seems to levy a royalty even upon mines, the rights to which have been legally alienated from the Crown already... It is illegal taxation, or rather confiscation... The courts of a free country may be relied upon to defeat any such formal violation of justice.[13]

The GMAPQ and its allies were successful in having the Mercier law repealed, but their experience also crystallized the necessity for broader communication within the mining industry. Robert W. Ells, an Honourary Member and esteemed geologist who worked for the GSC, put it this way:

---

[12] GMAPQ, Council Minutes, January 1891 to September 1893, p. 83-85, vol. 1, Call no. MG28, Accession no. I 394, National Archives of Canada, hereafter NAC.

[13] R.W. Raymond, "Notes on the Quebec Mining Law," *JGMAPQ*, vol. I (1891-2- 3): 16-21.

> If the members of the General Mining Association of Quebec do nothing more than arouse an interest in the proper and scientific development of our mineral resources, it would prove itself worthy of the thanks of the whole Dominion…
>
> The formation of local mining clubs, where items of interest in connection with the different branches of the industry might be discussed, and in the establishment of night schools for the men, and questions of sanitation, etc., are also important ones. These clubs could formulate their ideas in a series of papers to be presented at the regular meeting of the society, and thus a healthy and beneficial interchange of thoughts and sentiment among all the departments of mining work could be promoted with much profit to all interested.[14]

The Quebec association was, itself, a product of an earlier group called the Asbestos Club, formed in 1881. These men met informally in an upstairs hotel room in Black Lake, with a common mandate to "promote, apart from political principles, matters, of such nature as will benefit the asbestos business in general and the people interested in asbestos mining in Canada in particular."[15] The Asbestos Club did not disappear with the founding of the GMAPQ, but received a small grant to supplement its annual fee of five dollars. In 1901, there were forty-one names on the roster, and membership attracted men from such places as Vancouver, Halifax, Boston, New York, and Frankfort (Germany). John J. Penhale, an active member of both the Asbestos Club and the GMAPQ, would later become deeply involved in the Federated Canadian Mining Institute and its successor, the Canadian Mining Institute.[16]

Legislation was not the only concern of the GMAPQ members. Mine safety was another. Many felt that proper inspections, carried out by mining professionals, would promote safer working conditions in the mines. In the event of an accident, a mine inspector could even act as a witness on

---

[14] Robert W. Ells, "The Importance of a Knowledge of Geology and its Kindred Subjects to the Mining Engineer and Prospector," *JGMAPQ*, vol. I (1891-2-3): 104-116.

[15] The Asbestos Club, documents, April 1890, and March 30, 1901, John J. Penhale Fonds, Société des Archives Historiques de la Région de L'Amiante, Québec. "Encouragement aux mines," *Progrès de l'Est* [Sherbrooke], June 13, 1900. Robert H. Jones, *Asbestos and Asbestos, Their Properties, Occurrence and Use* (London, England: Crosby, Lockwood & Son, 1897).

[16] Alfred Lloyd Penhale, North Hatley, Que., November 14, 1996, interview with author.

behalf of the mine operator. In an editorial in the *Review*, Bell noted that in Quebec no official records were kept of mining accidents, and no inquiry or report was ever made by the "so-called Mining Service of the Government."[17] Cave-ins were relatively common occurrences, and the reason why few lives had been lost so far was only because the men happened to be above ground at a fortuitous moment. Bell believed that the Association would be remiss in its duty if it did not send a representative to bring the matter to the attention of the provincial government.

Between 1891 and 1895, the GMAPQ held twenty regular meetings, and seven special sessions to discuss a Quebec mineral exhibit for the Chicago World's Fair, the new licensing tax on powder magazines, the custom tariffs imposed on mining machinery, and the need for a national museum to house the collections of the Geological Survey of Canada. Members had also arranged ten field trips to collieries in Cape Breton, asbestos mines in Black Lake and Thetford, copper mines in Capleton, and iron works at Radnor Forges and Grand Piles.

In 1893, the GMAPQ hosted the International Mining Convention which was a resounding success. Held in Montreal, it took place over four days in February, and more than two hundred people had travelled from across North America to participate. The AIME's sixty-fourth annual meeting was pre-arranged to coincide with the Montreal event, and other participating groups included the Mining Society of Nova Scotia (MSNS), the Asbestos Club, and the Mining Association of Ontario. The program focused on Canadian mining laws, mining practice, and the mineral industries across the Dominion. The influence of the AIME was substantial, and the American delegates eagerly attended the Canadian sessions. During the opening reception in Windsor Hall, John Birkinbine, President of AIME, not only gave an overview of the work done by his institute, but mentioned similar organizations in other countries where engineers, like themselves, could share expertise and companionship.[18] His remarks show how much attitudes in the industry had changed in North America in just a few years:

These expeditions, through which we meet eye to eye, and voice to voice, our friendly competitors, to discuss the interests and the scientific aspects of the industry which absorb us, have been of great personal and national benefit. It is thus we learn how much has been accomplished by persistent and intelligent labor, how much remains to be achieved, and how

[17] B.T.A. Bell, Discussion, J.B. Smith, "Mine Inspection," *JGMAPQ*, vol. I (1891- 92-93): 31-37.

[18] John Birkinbine (1844-1915) of Philadelphia was President of AIME from 1891 to 1893. He worked as a consulting engineer in the iron and steel industry, and specialized in blast furnaces.

by free exchange of ideas and of production, friendly understanding is promoted and personal acquaintance is built up.[19]

Robert Adams, Vice President of the GMAPQ, was in whole-hearted agreement, and it is easy to see why these early meetings were so rewarding:

> The pleasure arising from association with men of kindred pursuits adds to the profit gained from their ideas; and the "flow of soul" at the banquet that follows the "feast of reason" at the convention unites in friendship those whose interest as well as delight it should be to work in harmony.[20]

Even though the organization of the Canadian mining industry was progressing on a provincial basis, several members who attended the GMAPQ's Fourth Annual Meeting suggested that the MSNS and other provincial mining groups should band together. Because several prominent members were strongly in favour, a committee was drawn to thoroughly study the matter. Whenever a major decision or strategy was required, the common practice was to form a committee to investigate the matter, and then report its findings to Council, who would then take appropriate action. This practice would be carried through to modern times.

For the fledgling Quebec association, there were more serious problems on the horizon. An exhausted Bell had tendered his resignation in the fall of 1893, claiming that his increased workload within the association interfered with the smooth running of the *Review*. In addition, he was also editing the *Canadian Mining, Iron, and Steel Manual,* which published yearly reports on the operation of Canadian mining and metallurgical enterprises, and he was also the Canadian correspondent for several British journals including the *Iron and Trades Coal Review* and the *Engineer Review*. GMAPQ Council members were divided on whether to accept his resignation. Some supported Bell's decision, and others feared that the association would collapse without him:

> I regret that Mr. Bell has taken this attitude. He has not only been the organizer of this Association, but a second Warwick - a king maker; appointing his presidents and offi-

[19] John Birkinbine, "The Development of Technical Societies," *JGMAPQ,* vol. I (1891-92-93): 179-191.

[20] Robert C. Adams, "Preface," *JGMAPQ,* vol. I (1891-92-93): iii-iv.

> cers. I can not see how we are going to run without him - except it be into the ground.[21]

In truth, B.T.A. Bell had been leading an exceptionally full life. By 1891, besides being Secretary of the Quebec association, he was Honourary Secretary of the MSNS, which a year later, would incorporate the Gold Miners Association and a smaller group known as the Coal Owners' Association of Cape Breton and Pictou counties.[22] In 1892, Bell had tried to encourage the formation of a provincial mining association in Ontario, but his efforts were more frustrating than fruitful. The difference was that in Nova Scotia, as in Quebec, mine operators had banded together when confronted by a harmful mining law. A similar rallying point was missing from the Ontario experience.

The Gold Miners Association

The Gold Miners Club was a somewhat informal group that met in a Halifax Hotel during the spring of 1887 to exchange opinions and expertise on improved mining methods, the recovery of gold during milling, and strategies for countering widespread pilferage by unscrupulous workers. On March 3, 1888, the Club evolved into the Gold Miners Association (GMA) whose objectives were:

> To mutually benefit and protect its members by facilitating the interchange of knowledge and ideas, by taking concerted action upon all matters affecting or relating to the gold mining industry of the Province, and generally, to promote said industry by all lawful and honourable means.[23]

According to Harry Piers, a gifted scholar who edited the *Journal of the Mining Society of Nova Scotia* from 1900-1915, the GMA was a curious compilation of the number 'three.' It was organized on the third of March, met on the third day of each month, and had three classes of members: active, or those directly engaged in gold mining or milling, associate,

[21] "Secretary's Resignation," *JGMAPQ*, vol. I (1891-92-93): 308-309.
[22] This may have been the same as the Nova Scotia Coal Mining Institute. See Harry Piers, "The Earlier Presidents of the Mining Society of Nova Scotia. With Some Account of the Society's History," paper presented at Mining Society of Nova Scotia Annual Meeting, June 1934: 1-16.
[23] Piers, ibid.

or those who did not qualify in the first group but who had an interest in the industry anyway, and finally, honourary members. Associates could not exceed one-third of the total number of active members, and in 1889, membership in the organization was thirty. Those who attended the GMA meetings over the next four years would praise the "hearty good-fellowship and the exuberance of all kinds of spirit and of wit, anecdote, and song."[24] Among the membership were several individuals who would later play important roles within the CIM: Dimock, Leckie, Libbey, and Hardman. Others, such as Benjamin C. Wilson and George W. Stuart, would become presidents of the MSNS. The sole honourary member was B.T.A. Bell.

Two very experienced men, from very different backgrounds, would also be prominent leaders in Nova Scotia's mining industry: Henry S. Poole and Robert Drummond. Poole was the Provincial Inspector of Mines for Nova Scotia in 1872, and became a staunch advocate of mine safety. A smooth talking and well-educated man, he oversaw the start-up and working conditions of mines in the province. He introduced the use of dynamite, and proposed new ways of utilizing and marketing coal. Poole later became the manager of the Acadia Colliery at Westville, and at the age of forty-two, was one of the most experienced mining men in eastern Canada. In 1892, he became the first president of the MSNS, and in 1898, was one of the Charter Members of CIM.

Robert Drummond's personality and expertise were very different from Poole's. Described as "a rough-hewn, feisty Scot with no formal training" he had immigrated to New York in 1864 before finally settling down in the collieries around Lingan, Nova Scotia.[25] In 1879, he had the unusual position of being both a bank-head boss for the Springhill Coal Company, and a mining correspondent for a Halifax newspaper. This gave him an inside edge. Concerned over the miners' drop in wages at a time when his company was experiencing profits, he saw an opportunity to support the miners by helping to found the Provincial Worker's Association. This was one of the earliest trade unions in North America, and for the next nineteen years, Drummond was its Grand-Secretary. Although he and Poole had common interests, such as mine education and mine safety, he was labour-oriented while Poole tried to solve the needs of owners and managers. It was only natural that the two men would clash.

[24] Ibid. Meetings of the GMA were held rather sporadically from 1887 until 1889. Pigot refers to it as "the secret society" because reports of its existence were scant in the pages of the *Halifax Critic*. It was not until the *Review* became the official organ of the Association that it gained wider acceptance and support. See David Pigot, *The Mining Society of Nova Scotia. 1887-1987* (Halifax: Mining Society of Nova Scotia, 1987), 125 p.

[25] Pigot, ibid.

There were many factors which led to the amalgamation of the independent mining groups into the new MSNS. Four months earlier, Poole had submitted an article to the *Halifax Critic* soliciting members for a Nova Scotia Coal Mining Institute. His actions were largely in retaliation to what he regarded as inept government action by the imposition of a royalty on coal. A group of twenty-two sympathetic men had already met in the offices of the Acadia Coal Company in Stellerton, and an open invitation for more members had been printed in the paper:

> The Institute supplies a want long felt, and enables managers, overmen, and others interested in mining, to periodically meet together and exchange their practical ideas and experiences.[25]

Those wishing to join were asked to communicate at once with Charles Fergie, in Westville. It was no coincidence that Fergie and Poole would later have positions of leadership within the more powerful MSNS, as both had gained valuable experience on the front lines of policy and industry.

In the spring of 1892, the rumours over an impending royalty on coal caused a shock wave throughout the industry. Coal miners, newspaper columnists, explosive manufacturers, and others, appealed to the GMA for support. After all, they were in the best position to approach the government on the issue, and many suspected that gold mining was next on the legislative agenda. Before taking the matter any further, the GMA Council thought it wise to add more weight to its numbers. It appointed a committee to study the formation of a united miners' association that would represent all aspects of the Nova Scotia's mining industry.

The Mining Society of Nova Scotia

On March 30, 1892, twenty men assembled at the Halifax Hotel to form the Mining Society of Nova Scotia, a close remake of the Gold Miners Association (GMA). Before they could tackle the royalty problem, the first order of business was to elect an executive which featured several of the GMA's most prominent officers. Poole was elected President, and two of the three Vice Presidents were Hardman and Leckie.[27] Harry McNab Wylde was named Secretary, and Wilson and Stuart, both past presidents of the GMA,

[26] Pigot, 23. Originally appeared in the *Halifax Critic*, November 27, 1892.

[27] The third Vice President was David McKeen, who later became Lieutenant Governor of Nova Scotia. J.R. Lithgow was Treasurer.

were elected Councillors. In May, the GMA's assets, debts, and membership list were officially transferred to the MSNS. It is important to note that this union represented the principal mining sectors of the province: Poole from coal, Hardman from gold, and Leckie from iron. In the future, every attempt would be made to alternate presidents from different sections of the industry, a tradition which would also be carried through the history of CIM. As Honourary Secretary, Bell watched over the proceedings with great interest. In the future, he would spend even more time on the train between his Montreal, Ottawa, and Halifax offices, undoubtedly complicating his family life.[28]

When Poole delivered his presidential address to the Society a year later, he credited the GMAPQ for leading the way in the fight to oppose unjust government legislation. He also attributed the words of Premier William Stevens Fielding as having had a positive effect on the society's achievements.[29] As Fielding was comfortably seated in the audience that night, how much of Poole's statement was purely political can never be known:

> The spark that at last lit the beacon possibly was that which fell from the Premier of the Province when he remarked to certain deputations, why do you not decide among yourselves as do other industries, and present your case as a united expression, then the government will, as far as possible, meet your wishes.[30]

The business of the MSNS since 1892 had proceeded along similar lines as its Quebec counterpart. Progress had been made in writing a constitution, strategies were developed for opposing royalties on coal, and many members were planning to attend the GMAPQ's International Mining Convention scheduled for Montreal the following year. Members also discussed Nova Scotia's contribution to a mineral exhibit for the Chicago World's Fair, the duty on mining machinery, and the controversial issue of professional certification for mining engineers. In addition, a seal, or logo, was chosen for the new Society. After several sketches were submitted, the ultimate decision was left in the hands of Poole and Hardman, both of whom favoured one which reminded them of the disbanded Gold Miners Association.[31]

---

[28] In 1899, Bell married Sydney McCarthy of Ottawa; their son, Gerald Gordon, was born two years later.

[29] William Stevens Fielding was Premier of Nova Scotia from 1884 to 1896, and the federal Minister of Finance, and Receiver General, in the Laurier government from 1896 to1911, and 1921 to 1925. He was elected an Honourary Member of CIM in 1899. *The Canadian Encyclopedia,* second edition, vol. II (Edmonton: Hurtig Publishers, 1988) p. 763.

[30] H.S. Poole, President's Address, Transactions of the MSNS, *Canadian Mining Review* (July 1891).

[31] The legacy of this design is readily apparent in the current seal of the CIM.

Aside from coal royalties, Poole had another favourite topic on which he had written a paper on "Legislation Affecting the Working and Regulation of Mines in Nova Scotia." Since a one-day meeting was planned in Montreal as a trial run for the 1893 mining convention, Hardman recommended that the Nova Scotia government send an observer. Hardman and Poole thought that this venue would be the perfect place to air some of the Society's grievances, and perhaps embarrass the government in front of other mining associations into changing its method of legislation. To everyone's surprise, including Poole's, Nova Scotia's Premier Fielding, who just happened to be in Montreal that day, appeared in the audience. A diplomatic exchange of views ensued, whereby Fielding defended his government's actions:

> We must deal with the truth that in former years laws were molded at the will of the mine owners; but at the present time... there are more than mine owners to be considered now, and we must speak plainly. The workingmen in the country have, in the past few years, made their voice heard and felt in the legislation of Nova Scotia; and in the desire to meet their reasonable wishes, some mistakes have been made.[32]

The representative of "the workingmen" was none other than Robert Drummond. As previously mentioned, he and Poole had been longtime opponents on mining-law reform, and Poole did not mince any words:

> The statute book is full of Acts hastily prepared and revised in the same laudable spirit, [and they] may be even amended by two and sometimes three separate Acts during the same session... without those chiefly affected having an opportunity of studying their bearing prior to their passage through the House... all Acts relating to mines should be prepared during recess and be subject to the criticism of all classes likely to be affected by them.[33]

The haphazard way surrounding the creation of acts, and their many amendments, coalesced in a heated discussion over the certification of min-

[32] W.S. Fielding, Discussion, "Legislation Affecting the Working and Regulation of Mines in Nova Scotia," *Transactions of the Mining Society of Nova Scotia* (1892-3): 36-37.

[33] H.S. Poole, Discussion, ibid., 30-31.

ing men. Fielding thought that Poole's observations about the workings of government were accurate, but that he had over-reacted on the certification issue, which was a means of ensuring a baseline of practical knowledge:

> A medical man is not necessarily a man of great skill. He gets his diploma and a big red seal, and he is sent out to cut our leg — if we are fortunate or unfortunate to fall into his hands... These certificates only prove that a man has received a certain measure of training, and if that man possesses brains and judgement, he should be qualified to do the requisite work.[34]

Poole countered by attempting to clarify his major complaints:

> I did not intend to question the efficiency of the certificate given to men who were examined. I took exception to the granting of certificates without an examination, and the leaving of the decision to men whom I did not consider were quite so competent to judge a man's fitness as those who were in charge of mines...[35]

In retrospect, perhaps Poole, Fielding, and Drummond were not that far apart in principle, and there is evidence that the Nova Scotia government was willing to listen to the needs and concerns of the industry. However, Bell was clearly on the side of Poole and the MSNS, which resulted in a series of articles in the *Review* criticizing those in Drummond's *Trades Journal* and Fraser's *Halifax Critic*. As no concensus could be reached on the certification issue, their arguments would last for several years. As for Poole and Drummond, so much hostility had been generated between them that it is hard to believe that the latter would eventually become a valued member of the MSNS.[36]

Mining People Organize in Ontario and British Columbia

B.T.A. Bell did not, in fact, resign from the GMAPQ as he had threatened, but had taken on the job of helping to organize two more mining associations: The Ontario Mining Institute (OMI), in 1894, and the British

[34] Fielding, ibid.
[35] Poole, ibid.
[36] Pigot, ibid.

Columbia Association of Mining Engineers in 1895. He served as secretary in the former, and honourary secretary-treasurer in the latter. In British Columbia, the flourishing association was reported to have sixty-five members, and was open to "professional mining engineers, geologists, metallurgists or chemists, or persons engaged in mining, metallurgy or metallurgical engineering." In a meeting in Vancouver in 1897, the association's members had expressed a "strong feeling in favor of affiliation to the Federated Canadian Mining Institute… it having been proposed by the latter body."[37]

In Ontario, the mining association had a rocky beginning. When the Provincial Mining Association of Ontario assembled in 1892, and again in 1894, Bell questioned its legitimacy claiming there were more lawyers, doctors, and speculators than actual mining men among its members . To counter this situation, he invited a more select group to attend a preliminary meeting in Toronto, which resulted in the formation of the OMI in 1894.[38] Among its charter members were Willet Miller, then a lecturer at the Kingston School of Mining, and Archibald Blue, Director of the Ontario Bureau of Mines. Both would make significant contributions to the CIM.

The year 1894 was truly pivotal. Bell's main objective had always been to form a national mining organization, and his efforts in this direction were about to take a step forward in the federation of Canada's two oldest mining societies in Quebec and Nova Scotia. Plans for affiliation had already been discussed at a previous meeting in Sydney, and the next occasion, scheduled for Sydney on July 9th, was destined to be a landmark event. As secretary, Bell had gone to great lengths to ensure that delegates and guests would travel in style:

> The fine passenger steamer *Bonavista* will sail from the Black Diamond Wharf, Montreal, at day break on Thursday, 5th of July, (passengers going on board the previous evening), reaching Sydney on the 9th… The Intercolonial Railway will furnish transportation to members and their guests from Levis to Sydney and return at a single fare… and the Canadian Pacific, Grand Trunk and other railways, will issue tickets at a rate of at least a fare and one-third… ample accommodation may be secured at [the New Sydney Hotel] at $2 per day.[39]

[37] "British Columbia Association of Mining Engineers," *Canadian Mining Review*, vol. XVI (1897): 140, 170, 237.

[38] "Annual Meeting of the Ontario Mining Institute," *Canadian Mining Review*, vol. XVI (1897): 170-1.

[39] Invitation, Secretary's Office, Ottawa, June 5, 1894, GMAPQ, vol. 2 (January 1894), NAC.

Towards the close of the meeting on July 12th, the two provincial associations met over dinner to discuss the business at hand. When Archibald Blue asked about the goals of the new federation, Bell confidently replied:

> Mainly the publication of a volume of Transactions, which would contain the papers read before all the societies in the organization. [It] would also place the mining men in a stronger position in such matters as the Dominion legislation.[40]

There was virtually no dissent at that time, as both groups had already accepted the idea in principle. Before the OMI was invited to join, Harry Wylde, Secretary of the MSNS, moved:

> That the Mining Society of Nova Scotia and the Mining Association of Quebec to hereby federate; and that a committee composed of the president and three members of the Society and the Association respectively, be appointed to meet and draw up a basis of federation.[41]

The name chosen for the new organization was the Federated Canadian Mining Institute. Despite the congenial atmosphere, everyone understood that the real work of the Federation still lay ahead.

The Federated Canadian Mining Institute

On November 6, 1894, a more serious debate over the feasibility of the new Federation took place in Halifax. After a good deal of reflection, the MSNS felt that the only positive aspect was that the Society could publish its Proceedings in the new *Journal of the Federated Canadian Mining Institute (the Journal)*. On the negative side, members were alarmed by the idea of a large governing Council, fearing the loss of their provincial autonomy and financial assets. At one point, Wylde was even asked to retract his initial motion to federate, and the entire venture was in danger of collapse. Still a dominant figure in the MSNS, Poole temporarily breached the impasse:

> It is desirable that there should be some organization with a Dominion character. I would not like to see the Mining

[40] Ontario Mining Institute, *Minutes,* vol. 1 (September 1894), p. 12, NAC.
[41] Pigot, 34.

> Society of Nova Scotia become of secondary importance, and I do not see how it can under this new formation. I think giving it a Dominion character would add dignity to it in the eyes of the public. We have allowed the mining business to be run by the politicians alone; now we all want to have a say in mining matters.[42]

From that point onward, amendments to the Federation's formal structure went back and forth between the various societies for approval. By March 1895, the MSNS had trimmed its own formula down to the last few details. With John Hardman as Chairman, and B.T.A. Bell seated beside him, consensus was finally reached between the two groups.

In January 1896, the Federated Canadian Mining Institute was officially formed in Montreal with Robert G. Leckie as President. Leckie understood the rationale for establishing a national association, and during his inaugural address, elaborated on the accomplishments of the young Federation:

> Around us here are met men from every province of the Dominion — Engineers of high technical training and extended experience, explorers, who have shown pluck and endurance in their tireless search for croppings; miners, who have developed the discoveries, and metallurgists, who have extracted the refined metals from the ore… Our Institute, as you know, is a federation of Provincial mining societies, which in their own limited spheres have been of great service to their members, both from a technical commercial and legislative point of view, besides promoting a spirit of social good fellowship.[43]

Sensing the diverse, and somewhat apprehensive mood, Leckie did as much as possible to put his audience at ease. He confirmed that the Federated Institute would carry the work of the individual provincial associations to a wider sphere, both geographically and topically. The associations, he urged, would still look after their immediate affairs which he believed to be of greater practicality. This made perfect sense because land titles, conditions and terms of mining leases, laws regulating the workings of mines and employment, tech-

[42] Ibid.
[43] Robert G. Leckie, "President's Address," *Journal of The Federated Canadian Mining Institute (JFCMI)*, vol. 2 (1897): 300-3.

nical education, and local taxation were all under provincial jurisdiction. The Federated Institute would devote itself to technical and scientific matters, and would encourage students who "came from an atmosphere of theory and the laboratory of scientific experiment, to rub shoulders with the hardy explorer and those engaged in the active pursuits of the profession."[44]

Under the auspices of the Federated Institute, the first Inter Provincial Congress of Canadian Mining Engineers, Mine Managers, and Mining Men was held thirteen months later, on February 5, 1897, in the Windsor Hotel. Many papers were presented by leading authorities of the day. At the dinner, the guest list was a veritable "Who's Who" of Canadian business, government, industry, and academia.[45] Lord Aberdeen, Governor- General and Patron of the Institute, was appropriately piped into the great banquet hall in the presence of more than 150 people. After dinner, Leckie gave the toast to the Queen, followed by the singing of "God Save the Queen." Glasses were raised to His Excellency the Governor General, and the audience responded with a few bars of "For He's a Jolly Good Fellow." More toasts honoured the House of Commons, the Local Legislature, the Mineral Industries of the Dominion, and Our Mining Schools, interspersed among formal speeches, and songs such as "The Minstrel Boy" and "The Brigadier." Last, but not least, B.T.A. Bell "sang most acceptably, 'Drill ye Tarriers, Drill,' with characteristic musical accompaniment, the fine chorus being taken up lustily by the company."[46]

Throughout the next year, Hardman and Leckie made heroic efforts to keep the fragile Federation together even though at this difficult period, both men were in the prime of their careers. Hardman was working as a consulting engineer in Montreal, and had just accepted a lectureship in mining at McGill University. In the interim, he had also courted and married B.T.A. Bell's sister-in-law. Leckie was in the process of leaving his position as general manager of the Londonderry Iron Company in Nova Scotia for consulting work in Sudbury. He also chaired a committee monitoring the Civil Engineer's Bill, a piece of unwelcome legislation that caused much distress among the mining community.

In the 1897 *Proceedings*, it was noted that the Federated Institute had published twenty-eight technical papers, and some gains had been made towards persuading the government to agree to a more liberal

[44] Ibid.

[45] The head table included W.S. Fielding; G.M. Dawson; the Hon. Sydney Fisher, Minister of Agriculture; the Hon. Thomas Chapais, Commissioner of Colonization and Mines for Quebec; Dr. Anderson, United States Consul General; Dr. Peterson, Principal of McGill University; and, John McKergow, President of the Montreal Board of Trade.

[46] "Annual Dinner," *JFCMI*, vol. 2 (1897): 312-337.

arrangement on machinery tariffs. As promised, the *Journal* carried the annual reports of affiliated societies, and a movement was afoot to establish "a scientific and industrial collection of minerals, reference library, and reading room of mining literature" in the Macdonald building on Montreal's Victoria Square. These visible successes were not enough to satisfy a growing unrest among the membership. Clearly, the survival of the Federation depended on a higher calling. They needed to attract professionals who were more concerned with education, development, and technology transfer than in constantly fighting various bits of unpopular legislation.

Ultimately, it was finances that caused the demise of the federation and its replacement by the Canadian Mining Institute. The truth was that the organization was suffering from a major lack of funds. The provincial mining associations simply did not contribute enough money to keep the Secretary's office afloat, and the annual fees paid by each group were disproportionate. The Federation executive decided to appeal to the Dominion government for a grant to assist with publications. With Hardman at the helm, a steering committee was appointed to study the matter further. Assisted by Wylde, secretary of the MSNS, Leckie, and Blakemore, Hardman concluded that the Federated Institute's annual expenses were higher than its revenues. Because publishing the *Journal* was so expensive, there was little left over for promoting the Canadian mineral industry. The only solution was to reorganize.

The Canadian Mining Institute

Although B.T.A. Bell had worked hard towards the creation of the Federated Institute, he believed strongly that:

> Canada as a mining country was too young for the establishment of an exclusive corporation, and he would favor a new organization along the broadest possible lines. With respect to the existing provincial organizations it was not intended to interfere with them in any way.[47]

A circular outlining the newly proposed constitution and by-laws was written and sent to the affiliated societies. The official name would be The Canadian Mining Institute, and its objectives were listed as follows:

[47] "Proceedings Annual Meeting," *JFCMI*, vol. III (1898): 191.

> To promote the Arts and Sciences connected with the economical production of valuable minerals and metals, by means of meetings for the reading and discussion of technical papers, and the subsequent distribution of such information... through the medium of publications. [To] establishment a central reference library and a headquarters. To take concerted action upon such matters as affect the mining and metallurgical industries of the Dominion of Canada [and] To encourage and promote these industries by all lawful and honourable means.[48]

Membership would include "persons engaged in the direction and operation of mines and metallurgical works, mining engineers, geologists, metallurgists or chemists, and other such persons as the Council may see fit to elect," along with students entering the mining or metallurgical engineering professions. In addition, honourary members would be those who had distinguished themselves by their "literary" or scientific achievements, or who had promoted the development of the mining industries in some other fashion.[49]

The first Annual Meeting, held in Montreal on March 4, 1898, had "a large and representative attendance of Canadian mining engineers and others interested in the mining industry." As Chairman, George E. Drummond briefly traced the recent developments, after which members analyzed, clause-by-clause, the newly drafted constitution and by-laws, and adopted them without further revision. Officers and Council were then elected. John Hardman would be President, and William Carlyle, George M. Dawson, Charles Fergie, and John Blue would serve as Vice Presidents. A.W. Stevenson continued as official Treasurer, and B.T.A. Bell, who must have been very pleased with the proceedings thus far, remained as Secretary. He read aloud the names of the twenty-three Charter Members, and those of sixty-three additional men, from GMAPQ, who transferred their memberships. The GMAPQ was then dissolved.[50] This brought the total roster to eighty-six, including the Institute's Patron, His Excellency Lord Aberdeen.

With much of the main business out of the way, the rest of the evening was devoted to administrative details. Bell was authorized to

---

[48] Canadian Mining Institute (CIM), *Council Minutes*, March 4, 1898, vol.1, NAC.

[49] The term "literary" could refer to anyone who had distinguished himself as a professional writer on mining, such as the role that B.T.A. Bell filled as editor of the *Review*.

[50] The Ontario and B.C. provincial mining associations remained as separate groups.

print copies of the Constitution and By-Laws, and membership roster, and to issue a circular inviting all Canadian mining men to join for an annual fee of ten dollars. A new seal was also adopted based on that of the Federated Canadian Mining Institute, and Bell was asked to design a membership certificate which would feature the large red seal.[51] A budget of fifteen hundred dollars was approved for the following year, and Bell was instructed to communicate with the Institute's Ottawa barristers to Dominion incorporation at the earliest possible date. This would ensure that individuals would not be held personally responsible for any debts incurred by the Institute.

Thus, the Federated Canadian Mining Institute was dissolved, and a new era began with the Canadian Mining Institute. It included members of the Quebec, Ontario, and British Columbia provincial associations, "Making a practically unbroken union from the Atlantic to the Pacific... and representing many working members and a vast amount of capital." [52]

The only dissenting voice of the evening was the MSNS, which chose to remain as a separate body. Although the reasons for this action are not well documented, there is an indication that some members of the MSNS were suspicious of national institutions in general, and feared losing their own hard-won authority to the "undue prominence of Ontario and Quebec."[53] There is also evidence that the MSNS wished to continue lobbying the Nova Scotia government, while the Institute was drifting further away from such encounters.[54] In particular, Archibald Blue, a Charter Member, was adamant that the Institute should not get involved in what he considered to be political matters:

> I desire to enter a protest against the practice on the part of our Institute of meddling with matters which must be regarded as purely political... It is a mistake to use our Institute to shape the trade policy of the Country, one way or another. I think we should avoid all questions which are calculated to create divisions along political lines...[55]

[51] These certificates were issued to newly elected members in the early years of the Institute. They were subsequently discontinued for a long period of time, then revived again in 1972.

[52] "Proceedings Annual Meeting," *JFCMI*, vol. III (1898): 208. "British Columbia Association of Mining Engineers," *Canadian Mining Review*, vol. XVI (1897): 140, 237.

[53] Pigot, 42-3, and Piers, 4.

[54] Specifically, the MSNS wanted to continue pushing the government for additions to the Free List of Mining Machinery, whereas the CIM held that it should refrain from interfering in matters of trade policy.

[55] Archibald Blue, Institute correspondence, *Council Minutes*, March 4, 1898, p. 47, vol. 2, NAC.

Even so, the MSNS decision must have been a bitter blow to both parties, who had earnestly tried to accept each other's viewpoint. As CIM President, Hardman was very concerned about leaving the doors of communication open. He wrote several conciliatory letters to the MSNS, in one of which he states:

> I strongly desire to represent to the Mining Society of Nova Scotia (with which Society I feel myself strongly identified) the complete absence of any intent, on the part of either the committee or of the Institute, to supplant in any way the provincial work of the Mining Society, or in any way to interfere with its autonomy…[the new Institute] has room for all the members of the Mining Society who feel that they are interested in the welfare of mining throughout the Dominion.[56]

The MSNS would not formally rejoin the institute until July 1918, and then only under special articles of affiliation, not outright amalgamation. Nevertheless, relations remained cordial. Bell, as Honourary Secretary of the MSNS, may have helped to bridge the gap, and the two societies continued to hold joint professional meetings, and to arrange field trips within each other's areas of jurisdiction.

Throughout that first important year, the Canadian Mining Institute struggled to generate a new look. Legal incorporation had been granted by an Act of Parliament on June 18, 1898, giving the Institute its Dominion Charter.[57] This landmark was realized through Bell's persistence. The Institute had also received a much-needed Parliamentary grant of one thousand dollars to assist with publishing and other expenses, and an office, library, and reading room had been rented at the Windsor Hotel in Montreal. The library contained over 125 books and reports, several large drawers of maps, and a long list of periodicals and proceedings from similar societies around the world. Membership had risen from to 86 to 190, and people could read about events affecting them and their industry in the Institute's *Proceedings*. This publication also carried many of the technical papers presented at meetings.[58]

---

[56] John Hardman to MSNS, correspondence, MSNS Annual Meeting, April 13, 1898, in Pigot, 42

[57] 61 Victoria, Chapter 96, Bill no. 116, An Act to Incorporate the Canadian Mining Institute, Court House Library, Ottawa, June 18, 1898.

[58] The original membership is commonly stated as sixty-three. This does not take the founding members into account, which raises the total membership, in 1898, to eighty-six.

The efforts of Canadian mining and metallurgical professionals in organizing a national Institute soon attracted the attention and interest of the international science and engineering community. Early in January, the Federated Institution of Mining Engineers in Great Britain requested status as a corresponding society, and proposed that all mining societies of the British Empire follow suit. Privileges included attending each other's meetings, and receiving each other's publications at a reduced rate. By May, the Incorporated London Chamber of Mines had invited the Institute to send a mineral collection to the prestigious Mining, Metallurgy, and Machinery Exhibition in London scheduled for August 1899, and the same exhibit was to travel to the Paris Exhibition in 1900. Without a doubt, visibility of this kind contributed to the Institute's recognition in Europe.

The Institute had also cultivated a good working relationship with AIME which enabled members to stay current with new mining and metallurgical technologies on either side of the border. Cooperation between the two societies has its own historical roots. From the mid-nineteenth century, there was a noticeable shift in the transfer of science and technology from Britain, as people increasingly looked to the United States for institutional affiliation and scientific exchange. Conditions which favoured this movement were proximity, shared geography, the easy cross-border movement of people, and a growing trade in commodities and manufactured goods. Although some Canadian scientists and engineers were members of British organizations and published in British journals, many others were seeking association with American peers and were sending articles to U.S. publications.

In the mining industry there were particularly close ties between Canada and the United States. Many mines had been developed with the aid of American capital, and were run by Americans, and much of the ore mined in Canada was smelted and refined in the U.S. In return, Canadians had much to contribute to AIME, and throughout its esteemed history, some CIM members would serve as officers of that organization. In addition to John Hardman and R.G. Leckie, who were elected vice presidents, James Douglas would be an AIME president in 1899. Born in Quebec, he was a graduate of Queen's College and studied at the University of Edinburgh. A mining and metallurgical engineer, he gained prominence in the copper industry in Arizona where he became president of the Phelps Dodge Copper Company. The James Douglas Medal (AIME) was established in his honour to recognize achievement in metallurgy. Others, such as Frank Dawson Adams, invited notable American scientists and scholars to the CIM annual general meetings, many of whom later became CIM members and served as valuable links to AIME.

The First Annual General Meeting of CIM was held in 1899 in the Club Room at the Windsor Hotel in Montreal at the beginning of March. During the business session, Bell presented the Report of Council for the past year. For the benefit of the new members, he began by explaining the similarities and differences between the old Institute and the new. Although there would be continuity in the membership, "only those in good standing shall be counted." Even though the Institute would continue to monitor legislation affecting the mineral industry, it would "refrain from expressing an official opinion on matters purely connected with trade or legislation, as the different provincial organizations still retained their identity, and were best fitted to deal with such questions." Finally, although the Institute was concerned about matters of education and professionalism, it was deemed unwise "to attempt to regulate the practice of the mining profession."[59]

Despite its best attempts to steer clear of legislative matters, the Institute's interest would soon be raised in three major areas: how to increase the efficiency of provincial mining bureaus, the problematic Civil Engineers Bill, and the necessity for establishing a Canadian Mint.

Provincial Mining Bureaus

The heated discussion over the Institute's relationship with the federal and provincial governments is an indication of the delicate balance between government and the private sector, and how the problems affecting the mining industry varied from one province to another. One issue centred on the lack of accurate mining statistics, and the lack of any recognizable standard for compiling them. This made it next to impossible for industrialists to gauge the value of Canada's mineral products. Although they agreed that some action should be taken, Institute members were concerned about the possible repercussions within their own jurisdictions. John Obalski, the Inspector of Mines for Quebec, thought that any positive suggestion made to his government would be welcomed. This was not the case in the Maritimes. Charles Fergie was concerned that "we have been driving the Provincial Government of Nova Scotia pretty hard for some years and they are beginning to think that we are pushing them a little too tight." W.H. Gallagher, a representative from British Columbia disagreed:

[59] "Annual Meeting," *Journal of the Canadian Mining Institute (JCMI)*, vol. II (March 1899): 198-9.

> If the Nova Scotia Government is doing its work the local Committee would not [be required] to act. The present Government of British Columbia has advertised throughout the Province for suggestions as to mining regulations, not only from the Mining Associations, but from anybody who wishes to offer them.[60]

Eugene Coste, an Ontario mining engineer who would later become a leader in the oil and gas industry, was quick to add that "the money spent by both the Dominion and local Governments in connection with mining has been so far thoroughly inadequate. The local Society might push the matter in any Province where it considered it necessary to do so." Perhaps Hardman was beginning to feel uneasy when he suggested that the topic be left to the discretion of Council, but not before accepting a resolution from the floor:

> That the Council of the Institute be authorized to instruct its Executive in the various Provinces to take whatever steps may be necessary to impress upon the Provincial governments the necessity of maintaining their Mining Bureaus in a high state of efficiency and the desirability of adopting a more uniform standard of compiling mining statistics.[61]

Civil Engineers Bill

Professional certification in some provinces had already caused much controversy during the preceding year and can best be described as a territorial dispute. The Canadian Society of Civil Engineers (CSCE) had been pressing the provincial governments for regulations on the practice of engineering, and the issue was soon reduced to one of definition. In the proposed legislation, the term civil engineer would include mechanical, mining, and electrical engineers. According to the CSCE, all of these disciplines were identical and should be subject to the same criteria. Curiously, when the legislation was being discussed in the Nova Scotia Assembly, none of the other engineers were represented. In the *Review*, Bell had chastised

[60] Ibid., 254.
[61] Ibid., 255.

the provincial legislature for passing a measure which he claimed was "an Act of egotism, arrogance, and blind pride [that] reflects on its originators, the Canadian Society of Civil Engineers."[62] Hardman's major complaint was not about giving a qualified engineer the proper status, but in imposing a tax of twenty dollars a year for the privilege. These funds were to be pocketed by the CSCE, a situation which he found outrageous. "Let the tax, if one must be paid, in the case of mining men, [go] into the treasury of the Nova Scotia Mining Society."[63]

The Federated Institute felt that it had already dealt with the situation in 1897 by adopting a resolution and forming a committee under the capable direction of Leckie:

> This society condemns clauses of the bill promoted by the Canadian Society of Civil Engineers, as far as they relate to the practice of mining engineering; further, that this society endorses the action of the Federated Canadian Mining Institute, taking action to defeat the measure in Nova Scotia and the other provinces of the Dominion.[64]

The Federated Institute successfully opposed the Civil Engineers Bill in Quebec and Nova Scotia by insisting that the word "mining" be deleted, but the issue resurfaced during the 1898 General Meeting. The CSCE had again promoted a bill, this time compelling every mining engineer and manager to be properly certified by passing a board examination. To do this, they would first be obliged to become CSCE members, something the CIM found totally unreasonable. It was not long before both parties were locked in a stalemate.

Although they had no intention of entering into the certification business, Institute members wanted a say in how it affected them, and the CSCE's Secretary, Professor McCleod, and John Hardman, finally confronted each other during an afternoon session. In what must have been a rather tense atmosphere, McCleod admitted that "immediately after it was found that the clause respecting mining engineers contained in the Bill was objectionable to the Mining Institute it was withdrawn."[65] Hardman responded by saying that "so far as we are concerned we did feel that

[62] "Amazing Arrogance," *Canadian Mining Review*, vol. XVI, no. 2 (February 1897): 37-38.
[63] Pigot, 40.
[64] "The Canadian Society of Civil Engineers' Bill," *JFCMI*, vol. 2 (1897): 349.
[65] "That Civil Engineers' Bill," *JFCMI*, vol. 3 (1898): 213.

any legislation governing the profession of mining engineering should emanate properly from an institute representing mining engineers, and not from a body of civil engineers."[66] Another CSCE representative expressed the hope that "instead of, as we saw in the papers, your opposing the Bill going through any other House, you will do all in your power to help us." To which Hardman replied, "Certainly, as long as you leave the mining engineers alone. I hope we will all move for the good of the profession." Bell thought that he had the last word on the matter: "It would have been better if the Canadian Society of Civil Engineers had sought our co-operation before."[67]

But the following year, the Civil Engineers Bill was again making headlines. Hardman was convinced that its persistence in the local legislatures was a preliminary attempt to pass a similar act in every province. He insisted that "a mining engineer is a mining engineer pure and simple, and he is not a hydraulic engineer or a sanitary engineer or the thousand and one things enumerated in this Bill." Left as is, the expression Civil Engineer meant practically anyone who acted or practised as an engineer:

> in advising on, in making measurements for, or in laying out, designing or supervising the construction of canals, aqueducts, drains, harbors, docks, lighthouses, river improvements, dykes, dams, railways, bridges, power plants, waterworks, sewerage works and all hydraulic, municipal, sanitary, electrical, mechanical or other engineering works, when the estimated cost of such work exceeds one thousand dollars.[68]

The new law would have made it mandatory for all engineers to join the CSCE or risk being fined fifty dollars in a police court. Had the Institute's conciliatory meeting with the CSCE Secretary been in vain? In an executive meeting, Council then passed another resolution to defeat the bill as it related to mining interests, and in April 1899, Eugene Coste appeared before the Legal Committee of the Ontario Legislature where he reported that the bill had been given the "twelve months' hoist."[69] The story of the Civil Engineers Bill was not over, just postponed for future debate. Discussions would continue in other provincial legislatures without resolution for several years.

[66] Ibid.
[67] Ibid.
[68] "Council Meeting," *JCMI*, vol. II (1899): 255.
[69] "Council Meeting," *JCMI*, vol. II (1899): 259.

A Canadian Mint

The last major issue on the agenda of the first Annual General Meeting was the necessity of establishing a Canadian Mint. Two members from British Columbia were adamant that action be taken:

> Whereas the establishment of a mint in Canada would be a great benefit commercially and otherwise. Be it resolved: That the President and the Secretary present a memorial on behalf of the Institute urging upon the Dominion Government the desirability of establishing a mint in British Columbia at the earliest possible opportunity.[70]

Bell thought that a Canadian mint was premature, and the use of gold for currency purposes was a backward step. In the *Review*, he later wrote that a nation's progress was measured by its trading ability, and the importance of gold coin lay in its international value: "When we cannot pay our foreign debts by the export of products, or the sale of securities, we may have to ship gold, and conversely when balances are due us from abroad we may import gold."[71] He suggested that the problem receive full consideration by Council as "so much can be said on both sides of the question that it would not be fair to the Institute to commit itself to this resolution without a full consideration..."[72] The postponement was not well accepted by British Columbia miners:

> We in Western Canada consider that a Canadian mint is an absolute necessity. The fact that there is $15,000,000 or more, of Canadian gold sent to the American mints annually should be at least a sufficient inducement to the members of this Institute who reside in the Eastern Provinces, especially in Montreal and Toronto, to support such a resolution. As well as being mining men I presume we are more or less commercial men, at least I am, and I can say that the sending of this bullion to San Francisco results in a great deal of business being done there and robs our own country of its legitimate trade.[73]

[70] These men were W.H. Gallagher of Vancouver, and J.D. Sword of Rossland. *JCMI*, vol. II (1899): 257.
[71] B.T.A. Bell, "The Proposal to Establish a Canadian Mint," *Canadian Mining Review*, vol. XVIII, no.6 (June 1899): 1.
[72] *JCMI*, vol. II (1899): 257.
[73] Ibid.

In the late nineteenth century, miners wishing to sell their gold had to send almost every ounce through the United States.[74] Transportation from remote mining sites was underdeveloped and expensive, and when shipping costs, insurance premiums, and handling by middlemen substantially reduced the miners' net return. Because an official government assay office did not exist in western Canada until July 1901, the selling price for gold was set at the lowest possible grade, and the market value was fixed. The miners knew that they needed a strong Canadian market for their gold, and their answer lay in the minting of gold and subsidiary silver coins to replace some of the paper currency. They also realized that in order for this to happen, there had to be a change in the political will.

The first official proposal for a mint was brought before the government of Sir John A. Macdonald in 1890 by a British Columbia senator who insisted that "there is no country worthy of the name of a nation that is without its own gold coin."[75] Canadian silver and copper coins, which had been in circulation since the 1870s, were supplied by the British Royal Mint. Business and political leaders viewed it as unprofitable and unnecessary to have one in Canada. The introduction of gold coins would supplant the country's successful paper denominations, and it was already a common practice to accept American eagles and double eagles as legal tender. Even concerns for a national identity failed to inspire Sir John A. MacDonald's Conservatives, who were more worried about their Party's waning support than in launching any potentially risky ventures.

The mint was also viewed in regional terms. Gold mining in British Columbia was not seen as economically important to either Ontario or Quebec, the centres of political power. As mining in Canada had not yet captured the imagination of the general population, people who might have influenced the political process were ambivalent about the issue. The proposal was soon overshadowed by a decline in world trade, which affected Canadian exports and temporarily halted immigration. However, changes would soon occur to reverse this trend, such as the election of a Liberal government under Sir Wilfred Laurier in 1896,[76] and the excitement over the Yukon gold rush two years later.

---

[74] James A. Haxby, *Striking Impressions: The Royal Canadian Mint and Canadian Coinage* (Ottawa: Royal Canadian Mint, 1984) p. 56.

[75] T.R. McInnes, in Haxby, 57.

[76] The Right Hon. Sir Wilfred Laurier, M.P., became an Institute Patron in 1899. He had personally been involved in the mining industry as an owner of the Laurier Mining Company of Black Lake, Quebec.

It is no coincidence that the desirability of a Canadian mint was being debated by CIM members the same year that the proposal was reintroduced at the 1899 session of Parliament.[77] One man who understood the situation completely was W.S. Fielding, who had recently been elected an Honourary Member. Initially not in favour, he would have been unable to ignore his mining friends for very long. They would have reminded him of the importance of the Klondike, which had already yielded placer gold worth in the neighbourhood of $2,500,000.[78] The Yukon had also been declared a separate territory, and Dawson City was the largest city west of Winnipeg and north of San Francisco.[79] The "idea of Klondike" intrigued the general public, and for those who could not seek fame and fortune for themselves, the gold rush offered an endless array of entertaining tales in the popular press. This served to heighten interest in mining in general, and soon placed the mint back on the political agenda.

The British government was not enthusiastic about this venture, and what eventually happened in negotiations between it and the Canadian government is a long story,[80] but during his campaign speech in 1900, Fielding gave the long-awaited announcement. He realized that if the mint could turn a profit during periods of least demand, then it might exist as a branch of the Royal Mint in London.[81] What he failed to make public at the time was its Ottawa location, which disappointed many western miners. Because agitation for the mint had begun with the British Columbia mining community, they were not satisfied until the Department of the Interior agreed to open a government assay office in Vancouver. At least they could now obtain an official analysis of their gold before selling it. This turned out to be a wise move because plans for the mint stalled for another six years. Canadians would not pocket their own gold coins until 1908.

---

[77] The proposal was reintroduced by W.W.B. McInnes, the son of T.R. McInnes.

[78] D.M. LeBourdais, *Metals and Men* (Toronto: McClelland & Stewart, 1957) p. 98.

[79] The Yukon became a Territory in 1898, separating it from British Columbia and the Northwest Territories. Until 1905, Alberta and Saskatchewan were part of the Northwest Territories. For one year, Dawson City was known as the "San Francisco of the North." The first Yukon census, in 1901, showed a total of 28,219 persons, most of whom lived in Dawson City. But the boom was short-lived, and within a decade, the population for the entire Yukon Territory had steadily declined to less than 5,000. Lebourdais, 104.

[80] For a brief history of the Canadian Mint see Haxby, 55-67.

[81] This could be accomplished if Canadian gold was used to coin British sovereigns worth £1, rather than using gold for domestic denominations of a dollar, leaving the existing metallic and paper currencies intact. The British Empire would therefore become a new market for Canadian gold.

The Institute Progresses

In March, it was again time to review the Institute's achievements at the Second Annual Meeting (AGM) in Montreal. Membership had increased to 297, which included seventeen from the United States, nine from Great Britain, and one each in China, Spain, and Hawaii. Twenty-three papers were presented, and these, together with discussions and proceedings, were published in the second volume of *The Journal of the Canadian Mining Institute (JCMI)*. Some of the best papers had even been reproduced in European and American technical and scientific journals, and the Institute library had grown by leaps and bounds. The Institute also succeeded in having gold-dredging machinery and appliances added to the "Free List" of importations for those working in British Columbia, and the Northwest and Yukon territories, and in Quebec, it endorsed the passing of a bill that exempted mining property from municipal taxation for another ten years.

Inquiries and advice concerning Canadian mineral resources and the mining industry were often directed to the Institute, and Bell was only too delighted to reply. He tried to take advantage of every opportunity to demonstrate the value of professional affiliations to his industry colleagues, and promoted the Institute beyond its Canadian borders. In fact, his activities were reminiscent of the days when he and Alfred Selwyn, then Director of the Geological Survey of Canada, had co-hosted the British Iron and Steel Institute, and the German Association of Iron Smelters, as officers of the old Gold Miners Association.

If CIM members wanted solid representation from western Canada, then the election of Samuel S. Fowler as President for 1900-01 was an excellent choice. He had served on Council since the beginning, and was prominent in the British Columbia mining industry. An American from Mount Vernon, New York, Fowler had graduated in engineering from Columbia University. He had moved to British Columbia when mining was in its initial stages, and worked for London and British Columbia Goldfields Limited. He would later play an important role in the development of the Kootenays. In Fowler's introductory address, he recognized the significance of residing in British Columbia and being chosen to head the Institute. The honour, he said, belonged to all of his colleagues and friends in that province who were involved in the mining business and wished to participate in its wider community. Although he alluded to their isolation and distance from Headquarters, Fowler insisted that the CIM could be made useful to everyone regardless of location, and that its work was dependant upon energetic and honest individuals:

> Every man must put forth his efforts to make the Institute a success. It does not do to sit by at the meetings, but we must do all we can to advance the mining interests of the country, and exercise ourselves largely as we did in British Columbia last summer.[82]

Bridging a far flung constituency, the summer meetings to which Fowler referred became important events in the history of the Institute, and predated the time when the AGM, and later, the District, Society and Division meetings were held across the country. Here, in a congenial atmosphere, members could engage in first-hand discussion of problems and successes, present papers, and participate in social functions hosted by local committees.

In 1899, some members from eastern Canada had journeyed by train to the mining districts of British Columbia. Although the return fare from Montreal by rail was just over $121, and the cost of hotel accommodations for three weeks was only $22.50, participation fell short of expectations. The delegates visited the collieries at Anthracite, the Hot Springs and National Park at Banff, the Le Roi, War Eagle, and Centre Star gold mines in the Rossland Camp, the Pelatin-Clerici process cyanide plant at Silica, the Silver King mine at Nelson, and the Canadian Smelting Works at Trail Landing. Official meetings, chaired by John Hardman, President of the Institute, were held in Rossland in the Miners' Union Hall and later, in Nelson. It was here that Fowler presented his popular paper on "The Ymir Mine and Mill." After other papers, and a lively technical discussion, members from British Columbia introduced a topic of much concern: The Eight Hour Labour Law. This piece of legislation illustrates how far apart the mining industry, government, and labour could stray on mining issues, and the willingness of the Institute to jump into the fray.[83]

An Eight-Hour Day

In small mining towns like Rossland, most workers belonged to the Western Federation of Miners, a powerful union which represented men from Mexico to Alaska. When James Wilks, the Rossland delegate to the Trades and Labour Congress of Canada, was elected vice-president of that organization,

[82] *JCMI*, vol. III (1900): 219.

[83] "Excursion to British Columbia," *Canadian Mining Review*, vol. XVIII (1899): 273. "Summer Meeting," *JCMI*, vol. III (1900):166-174. See also Jeremy Mouat, *Roaring Days: Rossland's Mines and the History of British Columbia* (Vancouver: U.B.C. Press, 1995), p. 88-91.

he insisted that the wider goal of the eight-hour day include "the Metalliferous Mines of British Columbia and Ontario."[84] Because of the growing strength of the labour party in British Columbia, coupled with an unstable provincial government, the Legislature accepted the union's demands. In 1899, an existing law was amended, bringing the eight-hour day to fruition.

As expected, the law did not serve everyone equally, and some mine owners, as well as workers, felt that the new rules infringed on their "freedom of contract." One major problem was that the government of British Columbia failed to recognize that the conditions under which underground miners were paid varied from one mining camp to another. Some miners feared that their salaries would be reduced, while at the same time, men employed in similar tasks, such as driving railway tunnels, could still work twelve to fourteen hours daily without penalty.

Mine owners in southeastern British Columbia enlisted the Institute's support to fight the legislation, and as a national organization, the Institute, despite its earlier avowal to stay away from legislative matters, felt obligated to respond since it had promised to pay close attention to mining issues throughout the Dominion. Indeed, the situation in British Columbia had become grave. Three months earlier, the Slocan mines had closed after workers took strike action against the owners. The miners' grievances were directly related to a loss of wages as a result of the shorter day, making this the first major dispute of its kind in the Kootenays. The climate at Rossland was not much better. Because the mine owners could not generate a profit, they had suspended operations in February 1900, under the guise of having to make repairs. Returning workers were then asked to accept a new contract system in lieu of a daily wage.[85]

At the summer meeting in Nelson, a resolution was introduced by two members from British Columbia in an attempt to clarify the situation:

> This Institute feels the necessity of recording its strong disapproval of recent legislation in this Province, legislation that was unasked for by the miners and which interferes with freedom of contract by restricting the hours of labour underground; the which has led to the closing down of many active mines in this province and has caused a disturbance of the cordial relations which had hitherto existed between capital and labour.[86]

[84] Mouat, 75.

[85] Rossland workers eventually went on strike from July 1901 to January 1902 over management's use of non-union members.

[86] "The Eight Hour Law," *JCMI*, vol. III, (1900): 171-173.

When Bell asked whether they wished the Institute to take immediate action, one of the members reluctantly admitted, "I do not know whether the local House would be much influenced by what the Institute thought," but they would certainly welcome whatever assistance the Institute could provide. His angry tone insisted that:

> The Government took no steps whatever to ascertain the views of the employers of labour, or to take the views of the miners throughout the Province, but considered it advisable to bring in a resolution and smuggle it through the last session of the House without anyone knowing about it.[87]

A general exchange then ensued over what role the Institute should play in such cases. A representative from Nova Scotia argued that the eight-hour law was a local issue, undeserving of the Institute's involvement. This would have been consistent with the organizational strength shown by the MSNS in dealing with its own affairs. British Columbia members were far more vulnerable to legislative actions because they lacked the support available to their colleagues living in more populated centres. Hardman, who often took the wider perspective, said "It is not interfering in local matters in British Columbia to have the Institute discuss a matter which is of vast importance to the Province's interests." B.T.A Bell was his usual articulate self:

> I think this is a question upon which the Institute may be fairly called upon to exert its influence. Any legislation which is so detrimental to the successful development of the mining industry… should be combatted by the mine owners and mine managers, not only of this Province but by their confreres in other sections of the country. The Institute, while partaking largely of a technical character, is… primarily a protective organization. It has been so in the past… and I think we should do all we can as an Institute to secure a repeal of this obnoxious interference with the mining industry of [that] Province.[88]

The resolution opposing the eight-hour day was forwarded to the Minister of Mines for British Columbia, but in spite of the Institute's best

[87] Ibid., 172. The clause at first referred to boys under the age of sixteen, but there was no mention of the eight-hour clause for adult miners.
[88] Ibid., 173.

efforts, the controversial law remained on the statute books. The real significance of this story, however, lies in the fact that the Institute held the concept of confrérie or brotherhood, above its original pledge to "refrain from expressing an official opinion on matters purely connected with trade or legislation."[89]

Of the many discussions during the second AGM in Montreal, in 1900, mining regulations took precedence over other issues. Institute members, many of whom were mine owners or managers, often dealt with a myriad of federal and provincial rulings which were commonly complex and inconsistent. Then, as in the decades to follow, what the mining community was actually looking for was a little stability and uniformity in the laws of the land. When an intended export duty on copper-nickel ores and mattes in Ontario, in 1899, would have harmed the nickel industry, the Institute successfully opposed the passage of the "ill-conceived" legislation.[90] The following year, when a proposed "Act to Amend the Mines Act" of Ontario would have imposed "absolutely ruinous and prohibitive taxes on the mining industry of the province," the Institute again took immediate action.[91] Even though it might have disagreed with the methods of the Ontario government, the Institute thought that at least the government was attempting to instill order.[92] This was certainly not the case in the Yukon, where regulations were often in turmoil.

The Yukon Royalty

According to Eugene Coste, the regulations governing placer gold mining were written by bureaucrats in the Department of the Interior in Ottawa, and approved by the current minister with seemingly little regard for their impact. Sometimes the clauses would change so frequently that even the Yukon Gold Commissioner could not keep track of them:

> One time a mining claim could be 500 ft. long; at another time, 250 ft.; and still, at another time, 100 ft; they extended to the rim rock at first, then to the base of the hill - both very uncertain boundaries, really impossible to determine with decision.[93]

[89] See footnote 64.
[90] R. G. Leckie, "No Nickel Export Duty," *Canadian Mining Review*, vol. XVIII (1899): 131-134. And "Nickel Export Duty; New Caledonia Supplants Sudbury," ibid., 211.
[91] H. Mortimer-Lamb, "Some Notes on the History and Recent Development of the Canadian Mining Institute," *JCMI*, vol. XIII (1910): 592. "Opposed to Export Duties," *Canadian Mining Review*, vol. XVII (1898): 36.
[92] J.M. Clark, "The Ontario Mining Law," *JCMI*, vol. III, (1900): 110-117, and "The Yukon Royality," ibid., 221-23.
[93] *JCMI*, vol. III (1900): 204.

Clearly, the bureaucrats had little understanding of working conditions in the Yukon, being far removed from this new mining territory. However, of far more concern to the federal government was the great number of American prospectors in the Yukon, and the large amount of gold leaving the country. The fear of "Manifest Destiny," or the territorial expansion of the United States as being inevitable, was still troublesome for Canadians into the twentieth century.

In an effort to gain more control, the Dominion Cabinet levied a royalty of ten percent on the output of gold from each claim in addition to the initial registration fee, and stipulated that every alternate claim on the placer grounds was to be reserved for the Crown. The new royality was enforced by a contingent of Northwest Mounted Police, who established a customs office at Lake Taglish, and police out-posts every fifty miles to Fort Selkirk. This not only maintained order, but helped establish a Canadian presence. The United States responded by dispatching its own troops to Alaska, and threatened to bar Canadians from trekking through Alaska. What could have become a nasty international incident was miraculously avoided, even though nearly all miners protested the new law regardless of their nationality.[94]

Comments made by Stewart Fowler during the 1899 Annual General Meeting indicate that the issue went far beyond the scope of economics:

> The prime lever of all legitimate success is honesty… I am not ashamed to say that I am an American, but I have seen more rascality in western mining towns than I have ever seen anywhere else. Some of that element has cropped up in the western part of Canada, and we should make it our business to put down that element and keep it down.[95]

Aside from urging its members to comply with the law, the Institute could do little except to register a complaint. The AIME Secretary, Rossiter W. Raymond, whom Bell praised as "the greatest living authority on mining law," clearly felt the royalities were too large, and the rules governing the operation of the Yukon mines were "unnecessarily multitudi-

[94] "Manifest Destiny" was used to justify the acquisition of California and the Oregon and Alaskan territories. See: Kenneth J. Kutz, *Klondike Gold. The Philatelic History of the Klondike Gold Rush* (Darien, Connecticut: Gold Fever Publishing, 1996) 181-200.

[95] *JCMI*, vol. III (1900): 220.

nous… The whole thing reminds one of a high-priced boarding school, where music and dancing and a pew in the church are, nevertheless, extras."[96] Joseph B. Tyrrell, a respected geologist who lived in Dawson City, asked the Institute to support a petition to the Minister of the Interior:

> That the present onerous royalty of 10 per cent on the gross quantity of the gold extracted from the placer mining claims in the Yukon district be reduced to 2 per cent, and that this royalty should be collected off the gold as it is taken out of the district, and at the same time, that correct statistics of the production of the country should be recorded.[97]

Unfortunately, Tyrrell's petition was not adopted, and the royalty remained in effect throughout the gold rush.[98] When production began to decrease sharply due to the depletion of the rich claims on Bonanza, Eldorado, Hunker and Dominion creeks, prospectors and miners began to leave Dawson City and the Yukon. As the importance of the royalty faded, the Institute turned its attention to more pressing issues.

Tyrrell was well-qualified to comment on the north, having explored parts of the vast region as a member of the Geological Survey of Canada from 1881 to 1889.[99] In 1901, in an interview with the Dawson *Daily News*, he predicted that Canada's mineral wealth might eventually equal or exceed that of the United States. At the time, he was quoted as saying "There were between one and three million square miles of country in Canada unprospected and the mineral wealth that lies hidden in this vast area will be an important factor in the progress of this northern continent."[100] With only fraction of the country surveyed, and without adequate supply routes, especially during the harsh winter months, most mineral deposits still awaited discovery. His prophesy would not approach reality until the use of aviation opened Canada's more northern regions to prospecting in the 1920s. He was right about the mineral wealth still to be found, but Canada's total mineral output would never surpass that of the United States.

---

[96] "The Yukon Mining Laws," *Canadian Mining Review*, vol. XVI, no. 8 (August 1897): 243-4.

[97] *JCMI*, vol. III (1900): 221.

[98] Kutz, 197.

[99] Tyrrell is best remembered for his major fossil discoveries in Alberta's Red Deer Valley, and the discovery of a large coal deposit near Drumheller, Alberta. The Royal Tyrrell Museum of Paleontology bears his name, as does the mineral tyrrellite. His wife, Edith Tyrrell, founded the Women's Association of the Mining Industry of Canada (WAMIC) in 1921.

[100] LeBourdais, 105.

The Geological Survey and the Mines Branch

The Geological Survey of Canada (GSC) was extremely important to the mining community and it applauded its efforts to expand the "practical" knowledge of Canada's geology. Founded in 1842 by William E. Logan (1798-1875), it remains the oldest scientific establishment in Canada. Logan directed its activities until 1869, after which Alfred R.C. Selwyn succeeded him.[101] Both men directed the Survey from field stations during the summer, as would George M. Dawson during his directorship. Because the western territories were largely uncharted, extraordinary survival skills and an eclectic scientific background were basic requirements for the Survey's officers. Selwyn took this one step further. He was particularly interested in upgrading the professional level of the GSC, and insisted on hiring mining geologists, topographers, and paleontologists for field and laboratory work. Unfortunately its remuneration for staff was "wholly inadequate to the services they rendered to the country." According to B.T.A. Bell, "it was no wonder that the best men were being allured elsewhere at considerable higher salaries."[102]

After Dawson's death in 1901, the Institute proposed Frank Dawson Adams for director of the GSC, but this idea was dismissed, probably by Adams himself. Robert Bell, who became the next Director, would clash with B.T.A. Bell many times over the necessity of creating a federal department of mines to address the specific interests of the mining industry in such areas as mineral production.[103] This was not the first time that the mining community had argued with the GSC. The Federated Canadian Mining Institute had once tried to persuade Dawson to hurry the GSC's preparation and distribution of geological reports. In a lively exchange, one member complained that "The prospectors start out early, about the time the snow gets off and the mosquitos get on, and they have not the opportunity of reading any of the reports until the next winter."[104] Dawson's apologetic reply spoke about overwork, understaffing, and the problem of reports from the GSC's field officers being due only in December.

---

[101] Morris Zaslow, *Reading the Rocks. The Story of the Geological Survey of Canada, 1842-1972* (Toronto: MacMillan, 1975). Sir William Edmond Logan (1798- 1875) was knighted in 1856. Alfred Selwyn was an Honourary Member of CIM. Many prominent CIM members worked for the GSC at some point in their careers but are too numerous to mention.

[102] Nominate Dr. Adams. Mining Men Want Him at the Head of the Geoogical Survey. *The Gazette*, Montreal, 7 March 1901.

[103] "The Two Bells Battle," in "B.T.A. Bell - Canadian Mining's Spokesman," *Canadian Mining Journal*, Centennial Series, vol. 99, no. 6 (June 1978): 29-30.

[104] Col. S.W. Ray, *JCFMI*, vol. II (1897): 308.

The cause was later taken up by Mortimer-Lamb, who succeeded B.T.A. Bell as Secretary, and the Institute can claim a pivotal role in the establishment of a new Department of Mines, on April 27, 1907.[105] Although Mortimer-Lamb had favoured a merger of the GSC with the new Department of Mines under its own minister, the appointment of Albert P. Low as Director in 1906 resulted in an internal reorganization of the Survey. The bill passed by the House of Commons described the Department of Mines as having a geological branch and a mines branch, with possibly a third administrative branch to follow.[106] This allowed the GSC to retain its separate identity, as well as its historic name, and ever since, the GSC and the Mines Branch (later called CANMET)[107] have operated in parallel.

A Montreal Mining Museum

In its early years, the GSC functioned as semi-private organization under government contract. Originally called the Geological and Natural History Survey of Canada, the GSC was headquartered in Montreal until its administrative offices and natural history museum were gradually moved to Ottawa in 1881. The museum had filled an important niche in Montreal's learned circles, and the decision to ship it to Ottawa met with considerable opposition, especially from the mining community. Its departure left a void that could not be totally filled by either the smaller museum of the Montreal Natural History Society, the collections of McGill University, or the few private holdings around the city. The situation was largely rectified with the opening of McGill's Redpath Museum in 1882, to which Sir William Dawson, Principal of McGill, donated his personal collection of geological and paleontological specimens.

Five years later, however, the Redpath Museum still did not satisfy the needs of the mining community. In 1897, members of the Federated Canadian Mining Institute devised an ambitious plan to combine a scientific and industrial collection of minerals, a reference library, a reading room, and a public record of local mining projects. The premises at the Macdonald Building on Victoria Square were deemed large enough for expansion, and the museum could share a portion of the Survey's geological specimens as there appeared to be some duplication. The Montreal Board of Trade and Corn Exhange supported the proposal, and the Quebec Government had already pledged an annual sum of $1,500. B.T.A. Bell was

[105] Zaslow, ibid., 255-6.
[106] A. Ignatieff, *A Canadian Research Heritage; 75 years of Federal Government Research in Minerals, Metals and Fuels* (Ottawa: Minister of Supply and Services Canada, 1981) p. 1-5.
[107] Canada Centre for Mineral and Energy Technology.

confident that, if approached, the federal government would match that contribution. R.G. Leckie was committed to the idea and promised that Council would take the proposal to Ottawa:

> What we want is a commercial museum right here in Montreal directly in touch with the capital of the country... If we could only get such an institution established business men of the city would make very good use of it.[108]

The Montreal mining museum never materialized, probably due to lack of funds, but the reference library and reading room became a vibrant part of CIM from 1898 to 1965. Financial problems also held back the progress of the GSC museum in Ottawa. Because the venue was far too small, overcrowding made it difficult to engage in much comparative work. George M. Dawson, and the mining community, pressed the government for a new fireproof building as the collections were improperly stored and rapidly deteriorating. As Director, and a Charter Member of the Institute, Albert P. Low and the mining community continued to press the government on the issue.[109] Three years after Reginald W. Brock succeeded him, the right combination of personalities and political will brought the museum project to fruition. The GSC's collections were finally housed in the new Victoria Memorial Museum, which opened in Ottawa in 1911.[110]

Summer Excursion of 1900

By all indications, the Institute's August excursion of 1900 was one of the year's highlights. The Institute had pooled resources with the MSNS, and had invited members of AIME on a field trip to thirteen mining centres from eastern Canada to Newfoundland. Several government grants were received to finance the transportation over the three-week period. Delegates toured the asbestos properties at Thetford Mines, Quebec, the gold and coal mining districts of Nova Scotia, and in Newfoundland, the Wabana iron ore mine on Bell Island. The latter shows that long before Newfoundland joined the Confederation in 1949, its mining community was not forgotten by its Canadian colleagues in the Institute.

[108] "A Mining Museum for Montreal," *JFCMI*, vol. II (1897): 298-9.
[109] B.T.A. Bell, "A National Museum," *Canadian Mining Review*, vol. XVI, no. 5 (May 1897): 177-8.
[110] The museum contained geological, botanical, zoological, and ethnological collections. It was renamed the National Museum of Canada in 1927; in 1968, it was split into the National Museum of Natural Sciences and the National Museum of Man. In 1990, the former was renamed the Canadian Museum of Nature, and the latter became the Canadian Museum of Civilization.

Besides the site visits, the delegates participated in stimulating discussion and social interaction. Various communities along the route hosted the group in lavish style, providing entertainment well into the evening. At Thetford:

> The whole town was in gala attire, the various private buildings being handsomely decorated with flags and bunting, and the St. Jean Baptiste Society, with their band, were also present to assist in welcoming distinguished visitors.[111]

In Quebec City, delegates feasted on "an English supper of sandwiches and ale" at the Chateau Frontenac, toured the old streets by electric rail, and boarded a government steamer to view points of interest along the St. Lawrence River, all to the tune of "sweet music" played by the Royal Artillery Band.

After the participants had transferred from the Quebec Central to the Intercolonial Railway for the onward journey to the Maritimes, there were not enough cars to accommodate everyone comfortably during the height of regular summer travel. A special train, consisting of sleepers and a dining car, some of which came from the United States, was pressed into service. It was referred to as "the Institute's train", and effectively solved the transportation problem but added to the Institute's travel expenses.

The Passing of Old Colleagues

By the time Stewart Fowler handed over the CIM presidency to Charles Fergie, in 1901, membership had risen to 323. The largest gains were in Quebec, Ontario, and British Columbia. The Institute was also proud that four of its members were serving "with the colours" in South Africa, including Thomas J. Brown, who had an illustrious career in the Nova Scotia iron and steel industry after his safe return. It was inevitable, however, that such a close-knit organization would begin to suffer some distressing losses. The first was James King, an original member of the GMAPQ, who had joined the Institute during its inaugural meeting. He owned the King asbestos mine in Thetford Mines, had represented Megantic in the Legislative Assembly of Quebec from 1892 to 1897, and had always taken an active interest in Institute affairs.[112] His colleagues found it hard to believe that he had drowned in a boating accident during the summer.

[111] Schabas, ibid., 29.

[112] *Thetford Mines à ciel ouvert: Historie d'une ville minière* (Thetford Mines, Quebec La Ville de Thetford Mines, 1994): p. 18, 72-3.

George M. Dawson's unexpected death also shocked the mining community. A Charter Member and Vice President of the Institute, he is still considered as "Canada's foremost scientific mind of the latter nineteenth century."[113] Dawson was an exceptional geologist, cartographer, and naturalist, and his personal motivation enabled him to overcome a childhood illness which had left him with chronically weak lungs and a hunched back. He had a warm and engaging personality, and tales of his frontier adventures could keep audiences enthralled for hours. Sadly, at the age of fifty-two, he contracted acute bronchitis on a Thursday night in February, and was gone by Saturday.

The First Institute Branches

In 1902, Council took a momentous decision to form the Institute's first three Branches at Sherbrooke, Kingston, and Nelson. These were logical places to begin as they were centres of mining activity and contained some of the most experienced mining people. In Quebec's Eastern Townships, George R. Smith, had previously served in the GMAPQ, and was elected Chairman of the Sherbrooke Branch. Sherbrooke was accessible to most people involved in both the flourishing asbestos and copper mining industries. In Kingston, the School of Mining and Agriculture at Queen's University had been in operation since 1893, after the Ontario government realized that a provincial mining school was essential for training its future scientists and engineers.[114] The school's faculty included William H. Merritt, one of the Institute's Founders, and Willet G. Miller, an original member of the Ontario Mining Institute and a future president of the Institute. Like Sherbrooke, Kingston was also situated close to an active mining district.

Similarly, Nelson was a centre of mining in British Columbia. The enthusiastic members of the Nelson Branch, which was formed during the Institute's summer meeting, were already keen to start a second group. Led by S.S. Fowler, and assisted by the young Selwyn G. Blaylock, Institute membership in the region had jumped fifty percent since 1901 when B.T.A. Bell had personally promoted the idea of Branches:

---

[113] "Earth to Earth. A Great Geologist Returns to the Bosom of the Mother He Loved so Well. The Death of G.M. Dawson, Director of the Geological Survey." *Ottawa Daily Free Press*, March 4, 1903. Lois Winslow-Spragg, *No Ordinary Man: George M. Dawson* (Toronto: Natural Heritage/Natural History Inc., 1993) p. 7. Joyce Barkhouse, *George Dawson, The Little Giant (Toronto: Natural Heritage/Natural History Inc., 1989).*

[114] Charles F. A. Hews, *A History of Mining Engineering at Queen's. The First 100 Years: 1893 to 1993* (Kingston: Department of Mining Engineering, Queen's University, 1993) p. 35.

> The Secretary reported that successful meetings had been held at Kingston and Sherbrooke, and branches to be known as the Eastern Ontario Section and Eastern Townships Branch had been organized at these points in accordance with the resolution passed at the annual meetings [in 1902].[115]

Bell had recommended that the term 'branch' be used to designate these local sections, and it was an idea that flourished. It was becoming as difficult for members to keep track of the Institute as it was for Headquarters to stay in touch with the members. Council realized that because of the enormous distances between centres, the majority of members would be unable to attend regular meetings. It worried that members might therefore become indifferent or disinterested. Even though the Branch concept would be taxing at times, its overall success was apparent from the beginning. Indeed, the Branches would soon supply a steady stream of candidates for Institute President in the years to follow.

The history of the Institute's Branches tends to parallel that of Canada's mining, metallurgical, and petroleum industries. With the birth and demise of one-industry towns, new Branches were formed as old ones folded. In the early days, Branch meetings often included a formal program of technical papers, and many, having already passed the first stage of peer-review, were published in the Institute's *Proceedings*. These gatherings often became the focus of social life in many communities, and their existence was applauded by local employers. The importance of the Branches to the Institute was evident in that they had direct representation on Council until the concept of Divisions was introduced after 1920.[116]

The Library Moves to Ottawa

From 1902 to 1904, much attention would be given to the location and space occupied by the Institute offices and library. The latter was overflowing with reference books on mining and metallurgy, magazines, papers, reports, maps, photographs, and various bits of correspondence. Because of the limited space at the Windsor Hotel, it was felt that a new location should be found. One idea was that the Institute might share a portion of the new building acquired by the CSCE on Dorchester Street, although perhaps

[115] CIM, *Council Minutes*, June 13, 1902, vol. 2, NAC.

[116] The term "Division" was first used to denote geographic areas similar to the later Districts. From 1940 onwards it was applied to the technical divisions.

relations were still too strained over the Civil Engineers Bill to accept the CSCE's generous offer. The matter carried some urgency because Hardman, who had supervised the library's holdings, was about to leave Canada on business. Bell worried that the Institute would not be able to find a suitable custodian in Hardman's absence, and made a motion to Council:

> Whereas the present accommodation provided for the Library of the Institute in the Windsor Hotel, Montreal, is wholly inadequate and unsuited to the growth of the Institute, and with the object of improving the efficiency of this Collection for the use of the members, be it resolved that the Council is authorized to remove the Collection of books, maps, photos, etc. to the premises adjoining the Secretary's office, in Orme's Hall, Wellington Street, Ottawa.[117]

The Council's first reaction was to consult its barristers to see if the By-laws allowed it the "power to act" without putting the matter to a general vote of the membership. Council found that it had sole control of Institute business during its term of office, and its conduct was only subject to review during the annual general meeting. Moving the library was not problematic as long as the Institute still listed a Montreal address.

As expected, not all members were in favour of the move. John B. Porter, a McGill professor who served on Council, wrote to Bell with strong objections. He believed that if the collection was taken to Ottawa, it would be of comparatively little use to the Montreal mining community. If one had to travel to Ottawa anyway, then why not consult the more extensive collections in the Parliamentary Library, or those at the Geological Survey. He also reminded Council that the Institute could have free use of a room at McGill University, where books could be stored, used, and cared for with very little expense. Porter insisted that if an Ottawa location was really what was wanted, then Dr. Haanel, Director of the new Department of Mines, had also offered quarters in his building.

The library issue was just the beginning of a series of changes that would disturb the Institute. The Windsor Hotel had been an ideal venue because it was convenient for out-of-town visitors. One of Montreal's finest, it was located close to the business community and was easily accessible from the Canadian Pacific Railway's Windsor Station. Thus, there were

[117] CIM, *Council Minutes*, 1903, p. 122, vol. 2, NAC.

indirect advantages that outweighed the financial outlay needed for additional space. Room number four at the Windsor Hotel was also the Institute's official home, even though Bell usually worked out of the *Canadian Mining Review* office in a building owned by Orme & Son, on Sparks Street in Ottawa. He had already asked Council for an increase in his budget to cover library expenditures, and was willing to take on the added responsibility of curator. Bell, of course, had a vested interest in maintaining the collection as he had personally solicited most of its holdings, and as editor of the *Review*, it was far more convenient to have his sources nearby. Council relented, and on May 16, 1903, a lease was signed between the Institute and Orme & Son, and Bell set up the library in a hall adjacent to his office.[118]

Bell Leaves the Mining Society of Nova Scotia

Although the library transfer can be seen as a victory for Bell, the past few years had been stressful and demanding. His friend, William Blakemore, remarked that "as Secretary of the various Mining Societies which he founded and fathered, [Bell] did many things which aroused antagonism. This was due largely to his deeper insight and his forceful nature."[119] Constantly under pressure from various sectors, the most serious concerned his rift with the MSNS, which began as an argument between the GSC and several Society members over a lack of adequate geological maps. Intentionally or not, Bell presented his Nova Scotia friends in a poor light in the pages of the *Review*, and they accused him of being an irresponsible editor. Hidden angers and resentments, which probably had been simmering for years, began to surface, and Bell could do very little to make amends. George W. Stuart, President of the Society, moved that "the existing arrangement with the *Canadian Mining Review* cease… and that the publication of our transactions be given to the local papers."[120] In one unanimous sweep, the MSNS cancelled its subscription to the *Review*, and established its own *Journal of the Mining Society of Nova Scotia*, with Harry Piers as editor. As of February 1902, Bell ceased to be Honourary Secretary of the MSNS. Surely this was a bitter blow to the man who had devoted so much of his energy to its success.

Like the MSNS, there is little doubt that without Bell's leadership, the Institute might not have evolved so quickly, or earned the respect so vis-

[118] *Canadian Mining Review*, vol. XXII, no. 6 (June 1903): 146.
[119] *Canadian Mining Review*, vol. XXIII, no. 3 (March 1904): 41.
[120] Pigot, 46-7.

ible for such a young organization. In an appreciation of his work, Blakemore would later write that Bell was "the life and soul of the whole movement."[121] However, his influence could be rather controlling, and he might have effected a Headquarters move to Ottawa had he lived long enough. After the disappointing events in Halifax, he concentrated on Institute business and the multiple tasks at hand. By 1903, membership had climbed to 435, and the elected executive officers and Council were fairly representative of most sectors of the Institute's membership. Canadian mining was about to experience phenomenal growth as the silver discoveries at Cobalt were imminent. During the summer, Bell had been appointed co-commissioner of the Treadgold Royal Commission on the Yukon gold mining industry, and he had spent several months assembling evidence and exhibits for his report. Unfortunately, it was an assignment he would never complete.

Death of B.T.A. Bell

Preoccupied with the business of the day, Bell accidentally stepped into an open elevator shaft in his building on Sparks Street in Ottawa on February 18, 1904, and fell the equivalent of two stories. He remained in a coma for eight days in the company of his wife, Sydney, their thirteen-year-old son, Gerald Gordon, and his friend John Hardman. The doctors refused to give up hope until the rapid onset of meningitis destroyed all possibility of his recovery. Bell died on March 1st, the day before the Annual General Meeting was to convene at the King Edward Hotel in Toronto. His death was reported by major Canadian and American newspapers.[122]

When Council received the news in Toronto, it they could hardly believe it, or grasp its significance. Bell's death came as such a shock that Council was unsure of how to proceed. The technical sessions of the AGM were allowed to continue, but the banquet and a field trip to Niagara Falls were cancelled, along with the Thursday afternoon session when Bell's funeral would take place. A hastily assembled group of Council members consisting of Eugene Coste, as President, J. Stevenson-Brown, and six others including Willet Miller, convened that same evening. Although it did not have a quorum, the group agreed to hold the meeting anyway, transact the necessary business, and submit a report for approval at the next full

[121] *Canadian Mining Review*, ibid.

[122] "Death of Mr. B.T.A. Bell," *The Globe* [Toronto], March 2, 1904, and *The Ottawa Journal*, March 1, 1904. "En Passant,"*Canadian Mining Review*, vol. XXIII, no. 2 (February 29, 1904): 22; Wm. Blakemore, "The Late B.T.A. Bell," ibid., vol. XXIII, no. 3 (March 1904): 40; "Biographical Notice of B.T.A. Bell," ibid., 37-38.

Council meeting, a practice borrowed from the by-laws of the AIME. In his absence, a letter was read from Hardman indicating that the immediate administration of the Institute would be handled by a "Management Committee" composed of Coste, Stevenson-Brown, Hardman, and G. S. Davidson, of Ottawa, who would act as temporary Secretary under their direction.

At the AGM's final business meeting, it was resolved that:

> In view of the services rendered by the late Mr. B.T.A. Bell to the Canadian Mining Institute and in commemoration of his memory, the Council of the Institute shall found a medal to be called the B.T.A. Bell Medal, which shall, in the discretion of the Council be... presented to such persons as may be deemed to have rendered signal service to the advancement of Mining and Metallurgy in Canada.[123]

Coste, Stevenson-Brown, and Leckie then left for the funeral bearing a wreath with the words "C.M.I., 1904."

Council in Disarray

What followed in the weeks and months ahead was a series of confusing events, and even the *Council Minutes* from this period contain an array of comments penciled into the margins. The Management Committee, which met in Ottawa on March 19, 1904, had divergent views on how to govern the Institute, and on the appointment of Davidson as interim Secretary. As no consensus could be reached, the matter was relegated to Council. When the next Council meeting was called on April 9th, in Toronto, instead of at Headquarters, only three members attended: Eugene Coste, Fritz Cirkel from Montreal, and Willet Miller, who agreed to write the minutes. This was the second time that Council had convened without a quorum, and Coste justified continuing the meeting by citing the same AIME by-law as before. A lengthy session was held despite the consequences, and two resolutions were passed affecting the finances as well as the daily operation of the Institute.

The Toronto affair was just the beginning of a power struggle which would last an entire year. Coste, who resided in Toronto, complained that he did not have the Institute minute book because it had not been delivered to him upon request. Davidson, who worked in Montreal and was apparently

[123] CIM, *Council Minutes*, March 5, 1904, p. 145, vol. 2, NAC.

following orders from Hardman and Stevenson-Brown, replied that until he had their written consent, Coste's request would be denied. Letters written to Council by Hardman, Stevenson-Brown, and Frank Dawson Adams were read outlining their objections to the Toronto meeting, to which Coste replied that their interpretation of the Institute by-laws was faulty. The validity of the Management Committee was then called into question, and Hardman and Stevenson-Brown, who were accused of exceeding their authority, were subsequently dismissed from their duties by Coste.

Move to Stanley Street

The next manoeuver was one which must have left the remaining Council members shaking their heads in disbelief. Because he felt that the selection of an Institute Secretary had been done without approval, and he personally favoured the application of William Blakemore, Coste fired Davidson and appointed himself as both President and Acting Secretary. In his newly empowered position, he moved that the Institute's library be returned to Montreal and installed in the suites owned by Fritz Cirkel at 80 Stanley Street. Cirkel, who had submitted his bid in writing and was present at the time, generously offered to include utilities and his services as a librarian for the sum of $400 annually.[124]

When Hardman and Stevenson-Brown were informed of the changes, they drafted an urgent letter to Council and petitioned another meeting for September 9th. Hardman had documented the events that had transpired since Bell's death, distributing them in advance. His first step was to ascertain if the new meeting would have the authority of Council, and when this motion was presented, John Blue supported it. Hardman explained that it was not his intention to dwell on personality conflicts, but he worried that the Toronto meeting would put the Institute's future in question and would lead to a schism between Ontario and the other provinces. He stated that although "this condition is not perhaps anarchy… it certainly is not lawful."

After a suitable discussion, the Toronto meeting of April 9th was declared invalid, and members struggled with the issue of who, exactly, was in charge. Coste was not present, having sent a letter of regret in care of his solicitor. Hardman was especially upset at this lack of respect, and directed his anger to George R. Smith, who had taken the chair:

[124] CIM, *Council Minutes*, April 9, 1904, p. 3, vol. 2, NAC.

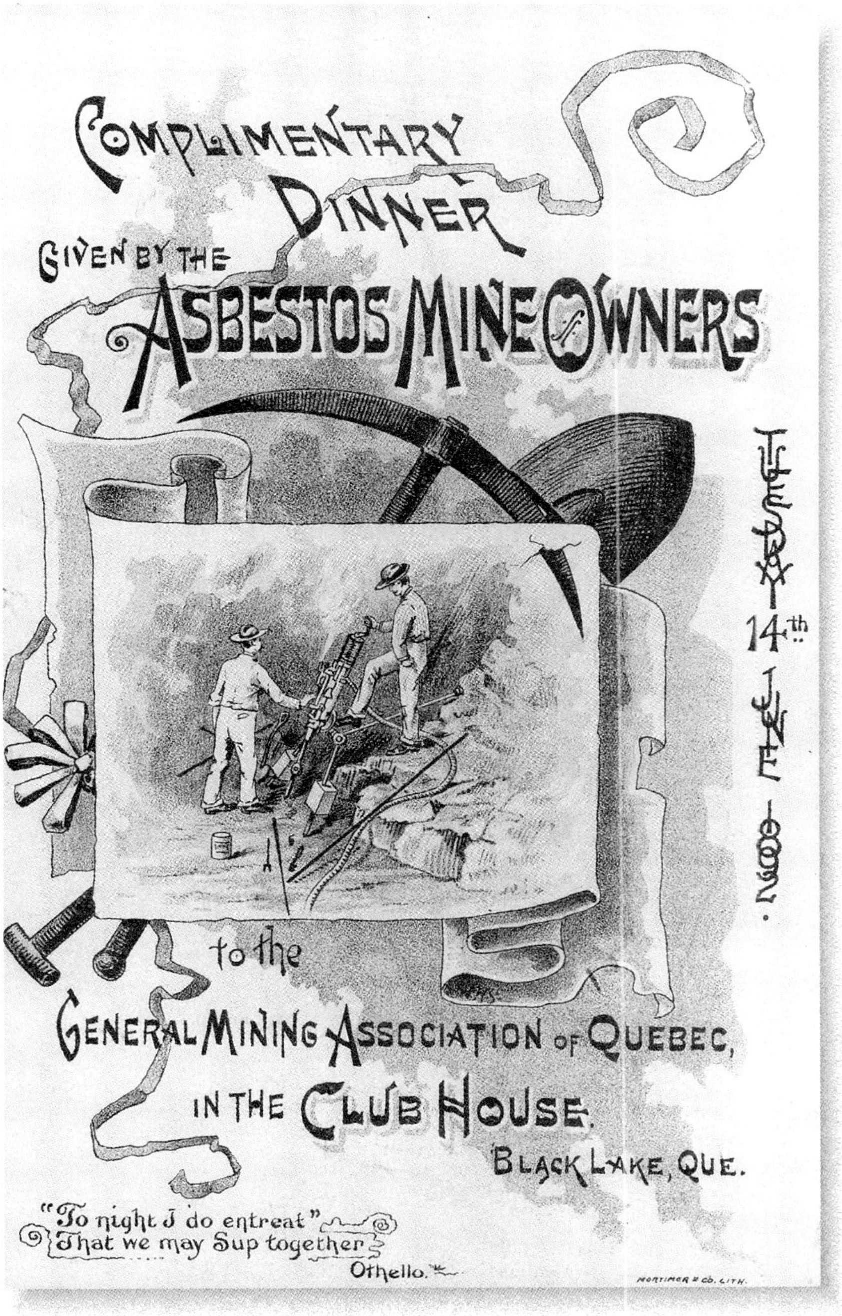
COMPLIMENTARY DINNER
GIVEN BY THE
ASBESTOS MINE OWNERS
to the
GENERAL MINING ASSOCIATION OF QUEBEC,
IN THE CLUB HOUSE.
BLACK LAKE, QUE.
TUESDAY 14th JUNE 1892.
"To night I do entreat"
That we may Sup together
Othello.
MORTIMER & CO. LITH.

Members of the General Mining Association of the Province of Quebec during the summer meeting on June 28, 1895. The group includes CIM Founders B.T.A. Bell (1); John Blue (2); George E. Drummond (3); J. Obalski (4); A.W. Stevenson (5); and J.J. Penhale (6). The Hon. George Irvine, Q.C. (7) was the Association's first President, in 1891.

Robert G. Leckie, first Chairman of the Federated Canadian Mining Institute, in 1896-97. A CIM Founder, he served on the first Council, and was a Vice President in 1905-06. He was also a Vice President of the American Institute of Mining Engineers in 1893-94, and President of the Mining Society of Nova Scotia in 1896-98. He has been called one of the fathers of mining engineering in Canada.

George E. Drummond, of the Canada Iron Furnace Co., was Chairman of the Federated Canadian Mining Institute in 1897-98. One of the 23 Founders of CIM, he was a Vice President in 1899-1900, and first Chairman of the Montreal Branch in 1908.

Many prominent members of the Institute were officers of the Geological Survey of Canada. Shown in this composite photograph of the staff in 1888 are: G.M. Dawson (7), CIM Founder; R.G. McConnell (8), CIM Founder; A.P. Low (17), CIM Founder; F.D. Adams (18), CIM President 1910-12; Eugène Coste (28), CIM President 1903-05; D.B. Dowling (29), CIM President 1918-19; and A.E. Barlow (39), CIM President 1912-14. (Geological Survey of Canada photo 97342)

George M. Dawson, Institute Founder and Vice President in 1898-99, is a legend in the history of Canadian geological exploration. He was third Director of the Geological Survey of Canada (1895-1901). (GSC photo 68773)

Right: Dawson (third from left) at Fort McLeod, B.C. in July 1879. (National Archives of Canada photo C18100)

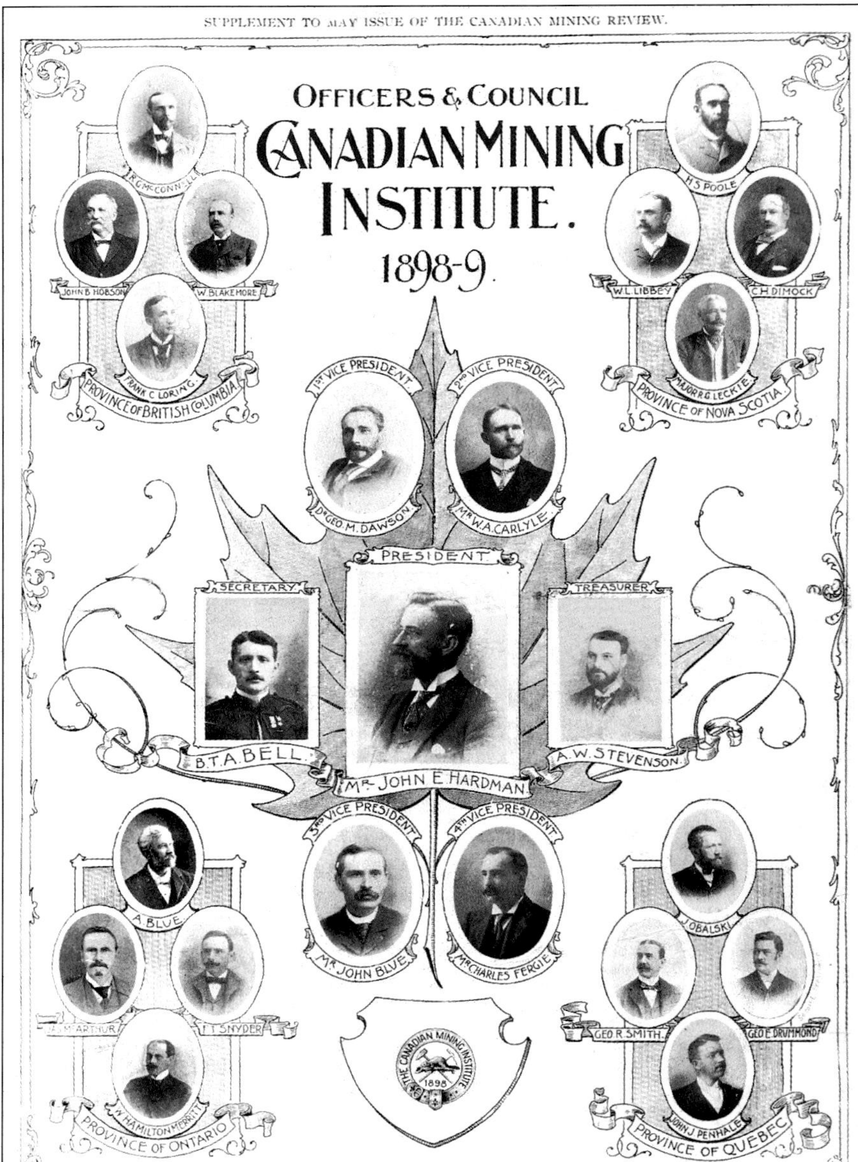
SUPPLEMENT TO MAY ISSUE OF THE CANADIAN MINING REVIEW.
OFFICERS & COUNCIL
CANADIAN MINING INSTITUTE.
1898-9.
JOHN B. HOBSON
W. BLAKEMORE
FRANK C. LORING
PROVINCE OF BRITISH COLUMBIA
H.S. POOLE
W.L. LIBBEY
C.H. DIMOCK
PROVINCE OF NOVA SCOTIA
1ST VICE PRESIDENT
DR GEO. M. DAWSON
2ND VICE PRESIDENT
MR W.A. CARLYLE
SECRETARY
B.T.A. BELL
PRESIDENT
MR. JOHN E. HARDMAN
TREASURER
A.W. STEVENSON
3RD VICE PRESIDENT
MR. JOHN BLUE
4TH VICE PRESIDENT
MR CHARLES FERGIE
A. BLUE
T.T. SNYDER
W. HAMILTON MERRITT
PROVINCE OF ONTARIO
THE CANADIAN MINING INSTITUTE
1898
J. OBALSKI
GEO. R. SMITH
GEO. E. DRUMMOND
JOHN J. PENHALE
PROVINCE OF QUEBEC

John E. Hardman, CIM's first President, and a tower of strength in its early days. A distinguished mining engineer, he had been President of the Mining Society of Nova Scotia in 1894-95.

Benjamin Taylor A. Bell, CIM's esteemed first Secretary and Founder. His vision led to the organization of Canada's first provincial mining associations, their federation, and their subsequent amalgamation as the Canadian Mining Institute. He was the guiding spirit of CIM during its formative years from 1898 to 1904.

> We cannot be dismissed by the president in consequence of what you just declared to be an illegal meeting. It does not concern me in the slightest, but as Mr. Brown is treasurer and custodian of the funds, it certainly concerns him a great deal.[125]

The Council members agreed that the original Management Committee, as appointed on March 15th, would stand, but that in Coste's absence, Hardman and Stevenson-Brown could constitute a voting majority. The minutes would then be mailed to Council, who would be asked for their opinion. If everyone was in agreement, the Committee would retain control. Hardman was confidant that this was the best way to untangle the mess, but was fully aware that "Mr. Brown and myself are going to have a monkey and a parrot sort of time [as] Mr. Coste is now Acting Secretary under his own appointment and he will not relinquish it at [our] request."[126] In the meantime, Stevenson-Brown was instructed not to release any Institute funds.

Drawing on his experience as a politician and a diplomat, Smith, who was the president elect, put the facts in perspective:

> as practical men… they were here trying to guard the interests of the Institute and from what I have heard since, I believe they felt it was to be regretted… [However] the Institute has incurred certain liabilities…[and these] must be taken care of and business must not be stopped for some little quarrel or some little difference between one member and the other… Unfortunately and to the great misfortune of the Institute, we have lost almost a secretary and a president and a treasurer combined.[127]

This sentiment was also expressed by Robert G. Leckie, who wrote that "The unfortunate condition of the Institute affairs is much to be deplored, and compels us now to appreciate the marvelous ability and tact constantly shown by our ever-cherished friend B.T.A." Leckie was aghast by the recent developments, and thought that many of the problems could have been avoided had an acting secretary been appointed immediately

[125] CIM, *Council Minutes*, vol. 2, September 9, 1904, p. 23, NAC.
[126] Ibid., 24.
[127] Ibid., 34.

during the Annual General Meeting. He had even consulted his own solicitor on the matter, whose advice was that "he could not see how the offices of president and secretary could be combined in the same person."[128]

Hardman, whose integrity and experience were the most compromised, was also the most vocal:

> Things have gone on regularly and undisturbed for seven years. There has never been any trouble before. We have had presidents from Quebec, Nova Scotia and British Columbia, and every one of them regarded the position... as an honorary [one]. The present president arrogates to himself, not merely the distinction of being the presiding officer of the Institute, but the chief executive officer, and if any subordinate does not please, he assumes the right to depose him.[129]

Despite all the objections, Coste remained as President and Acting Secretary until the next annual meeting, and the argument over the missing minute book persisted well into December. It was as if the book symbolized Bell's authority, and its fragile pages were the Institute's only tangible artifact. Sides were drawn again as Willet Miller supported Coste's view that the book had been purposely hidden by the Treasurer. Frank D. Adams strongly disagreed, saying that Stevenson-Brown had simply put it in his vault for convenience and there was no harm intended. The proper place for the book, he insisted, was at Headquarters, where it would be accessible to the membership. The meeting evolved into a one-issue debate until finally, a motion was passed to have the minute book deposited in the Institute's library, now located at Cirkel's offices in Montreal.

H. Mortimer-Lamb, Institute Secretary

At the Institute's Annual General Meeting on March 28, 1905, Council adopted a new slate of officers. Although the past year had left its mark on Council, never again would the Institute's leadership be challenged. The days ahead seemed full of promise and renewed energy. A new Secretary had also been welcomed into the circle, and he had

[128] R.G. Leckie, letter to J.E. Hardman, May 12, 1904, Sudbury, Ont., vol. 2, NAC.
[129] CIM, *Council Minutes*, September 9, 1904, p. 35, vol. 2, NAC

already won the admiration of the new executive. Harold Mortimer-Lamb was a journalist who had immigrated from England in 1889, settling in British Columbia. His working experience included an editorship of the *B.C. Mining Record* in 1897, publicity work for the British Columbia Department of Mines, and Secretary of the British Columbia Provincial Mining Association, which he helped organize. He also had been a member of the Institute since 1901. Alfred E Barlow would later claim that Mortimer-Lamb was "the greatest editor of a technical publication in Canada."[130] Along with having a similar background to Bell's, he was also completely neutral with regard to the various personalities on Council.

Coste did not attend the Annual General Meeting, but sent a strongly worded telegram asking that the Institute library remain on Stanley Street, and suggesting that the printing of the *Transactions* remain in Toronto "on the grounds of efficiency and economy."[131] To the Secretary's dismay, Coste had also neglected to include the *Council minutes* from February and March in the minute book, which he had finally obtained on principle during his last three months in office. Although Council's reaction is not fully detailed in the minutes, it was moved by Stevenson-Brown and seconded by Barlow that Coste be asked to relinquish all books, documents, and other Institute property in his possession, and not to incur any expenses beyond his final meeting.

As the newly installed president, George R. Smith suggested that a committee be appointed to relocate the Institute offices and library, and Mortimer-Lamb was asked to notify Fritz Cirkel that new premises would be sought by the end of April. Because of the numerous items on the agenda, the appointment of a Library Committee was postponed until the next meeting, as was a decision on printing. However, a new committee composed of Adams, Hardman, and Mortimer-Lamb was appointed to deal with editing and publishing. The evening concluded with the formation of yet another committee to compare the Institute's original By-laws with those recently drafted by Coste, who remained convinced that he had acted in good faith. Over the next few years, various amendments to the Institute By-laws would tackle everything from the direction of Branches, to the definition of members.

---

[130] "Proceedings of Sixteenth Annual Meeting," *Canadian Mining Institute Bulletin* (Bulletin), no. 25 (April1914): 64.

[131] CIM, *Council Minutes*, March 28, 1905, vol. 2, NAC. *Transactions* refers to the *Transactions of the Canadian Mining Institute*.

Another Headquarters Move

On October 6, 1905, Council approved the relocation of the offices and library from Stanley Street to suites occupied by the Canadian Society of Civil Engineers (CSCE) at 877 Dorchester Street. This was the same site that had been considered three years earlier, and was now deemed acceptable because "the possible rivalries and disagreements between the two societies are now matters of ancient history." Although the space was adequate, in 1908 Alfred E. Barlow announced that the Institute's Headquarters would likely be moved to Ottawa if two-thirds of the membership approved by letter ballot. His rationale was that the Institute's printing was now done in Ottawa, and the Secretary made frequent and prolonged visits there to carry out its business. The real reason, and one which prompted much debate, was that the Institute had already received a substantial grant from the Federal government, and more money was about to be requested for special projects. Along with satisfying many civil servants, the move would be an opportunity for a national organization to showcase its Headquarters.

Hardman opposed the relocation because he thought that the Institute would fall under the influence of the Department of Mines and lose its autonomy. He argued that because Montreal was a vibrant business centre with many mining industry head offices, many members could combine their work duties with Institute affairs. It was also a city easily reached by rail and ship, and the extra effort needed to travel to Ottawa would mean that some members might forgo serving on Council. This could result in the Institute being controlled by fewer men, a situation to be avoided at all costs.

A Return to the Windsor

Had the results from the letter ballot been successful, the Institute would probably have moved to Ottawa. The problem was that the ballot returns were dismal. Rather than base the decision on the wishes of a minority, Council formed a Committee to study the situation. Then, in November 1909, the library and Secretary's office were moved back to the Windsor Hotel. Within two years, these accommodations would again be deemed "wholly inadequate", and a committee consisting of Fergie, Smith, Barlow, Stevenson-Brown, Mortimer-Lamb, and several other members of the Montreal Branch would look at purchasing a building.

In Search of Permanent Quarters

Since 1910, the Montreal Branch had been campaigning for a Headquarters that would be "as unlike offices as possible, and rendered really attractive" to visitors.[132] Its suggestion was to purchase a three-storey house, and present it as a gift to the Institute. After careful study, the committee estimated that it would cost about $60,000 to own and maintain a building. When Willet Miller reported that the AIME "had been severely handicapped financially in the consequence of their efforts in this direction,"[133] Council decided to take a three-year lease at the Ritz Carleton Hotel, starting on February 1, 1914, after notice had been given by the Windsor Hotel that the Institute's rooms had been placed on a monthly rental system. The new location would suffice until 1917, after which the Institute would move again, this time to the Drummond Building on the corner of St. Catherine and Peel, where it would remain for the next sixty years.

New Branches

The Institute was now in a period of rapid expansion. At 850 members, projections for 1910 were close to 1,000. With six Branches in existence and more waiting for approval, much more thought had to be invested into structure and organization. During his term as president, Coste had "enthusiastically advocated" the establishment of Branches, and in 1906, became the first Chairman of the Toronto Branch. In 1907, Arthur A. Cole formed an important Branch at Cobalt, Ontario, and in 1908, George E. Drummond did likewise in Montreal. This resulted in an increase in the publication of technical papers, and the *JCMI* was now a tome of over 600 pages. Council felt that few of these technical papers would have been written had the Institute not existed in the first place, and recognized that local groups were far more effective in soliciting contributions than the Secretary.

The Branches were showing a lot of potential, but certain problems needed to be dealt with. Addressing the Council at the eleventh Annual General Meeting, Arthur Cole stated:

> We have great ideas in our Cobalt Branch, and we would like to receive support to carry them into effect… you can depend on the Cobalt Branch to stand heartily by the Insti-

[132] Ibid.
[133] CIM, *Council Minutes*, September 18, 1913, p. 91, vol. 3, NAC.

> tute, and you need not be afraid that we will start any agitation that will tend towards separation… we do, however, want something in the way of definite support and encouragement, and now is the time we most need it.[134]

Summer Excursion of 1908

In 1906, the Iron and Steel Institute of Great Britain had to decline an invitation from Council to hold a joint technical meeting in Canada. During the following year, a more ambitious plan was made to organize a cross-Canada tour of mining and metallurgical centres and to invite, not only the Iron and Steel Institute, but other societies from Great Britain and Europe to participate. The summer excursion of 1908 attracted some twenty-five overseas delegates who were joined at various points by several hundred CIM members, guests and invited officials.

The hosts of the "pilgrimage" which took place between August 24th and October 2nd were Fergie, Coste, Brock, and Miller, the Institute's President. The excursion was organized into three stages. In the first stage, the participants visited collieries of the Dominion Coal Company, the Nova Scotia Steel and Coal Company, and Dominion Iron and Steel Company works in Cape Breton, and the principal asbestos mines in the Thetford area of Quebec's Eastern Townships. Stage two brought people to the silver mines at Cobalt, and to the Moose Mountain iron mine, the mines and smelter of the Canadian Copper Company, and the Mond Nickel Company smelter in the Sudbury district. Stage three was a comprehensive tour of Alberta and British Columbia. Included were the coal mines in the Crowsnest Pass and at Nanaimo, and the gas wells at Medicine Hat. In the booming Kootenays, delegates visited the mines in Rossland, the Trail Smelter of the Consolidated Mining and Smelting Company, the mines and smelter of the British Columbia Copper Company at Greenwood, and the Granby Consolidated Company at Phoenix and Grand Forks. Besides the serious business, the Summer Excursion also took in cities and tourist attractions across Canada.

Mortimer-Lamb, who travelled the entire distance, credited the successful trip to the incomparable service provided by the railways, and the enthusiastic reception at local Branches, who made the delegates feel particularly welcome. In a special, hardbound souvenir issue of the *Bulletin*, Mortimer-Lamb chronicled the excursion with generous, and often

[134] "Proceedings of Twelfth Annual Meeting," *JCMI*, vol. XIII (1910): 21.

humourous detail, and added many of his own photographs. The volume also contained many of the speeches and papers presented during the tour, as well as discussions. The Institute of Mining and Metallurgy was so taken with the excursion that it offered a cash prize for the best paper submitted to the 1909 *Transactions* by a junior member of CIM. Its purpose was two-fold: to encourage younger members to write papers on mining and metallurgical topics, and to show appreciation for "the friendly feeling engendered and the delightful hospitality experienced at the hands of the members of your society."[135] Not until the Empire Mining and Metallurgical Congress of 1927 would the Institute again organize a cross-Canada tour of such proportions and extravagance.

The Institute and Other Professional Societies

Council had already been hard at work revising the By-laws since 1908. It had first made a careful study of the rules governing technical and learned societies in the United States and Great Britain, and had produced a series of amendments that would best serve its Canadian constituency. AIME, which predated both the Canadian and British institutions by more than twenty years, was clearly its guiding model. The problem was that Canadians were still very British in their thinking, and the AIME by-laws served an American audience. It was also a much larger organization. In 1904, its membership stood at over 4,000. Then, in 1911, it began a period of decentralization, with new technical divisions gradually assuming more responsibility for self-government. Although membership was not confined to U.S. nationals, it was largely restricted to professionals with academic credentials.[136]

In comparison, the Institution of Mining and Metallurgy (IMM) was only six years older than the Canadian Mining Institute, and although it had also learned a great deal from the AIME, the British were annoyed at having to acknowledge that the Americans were first.[137] Until the 1880s, mining in England was regarded as a necessary evil. Largely associated with coal, its long record of terrible accidents, worker abuse and fraud had been well-ingrained into public memory. Because of the perception that

[135] Called the Commans, Frecheville, and Marriott Prize, a one-time sum of $250 was awarded, divided between H.H. Yuill for "The Hosmer Mines, BC," and Frank B. Lathe for "The Granby Smelter."

[136] "The AIME: A Brief History (The Rand Aquamarine)" *Matrix: A Journal of the History of Minerals*, vol. 4, no. 2 (Summer 1995): 62-65. See also *Centennial History of the American Institute of Mining, Metallurgical, and Petroleum Engineers: 1871-1970* (New York: AIME, 1971).

[137] A.J. Wilson, *The Professionals.The Institution of Mining and Metallurgy 1892-1992* (London: IMM, 1992) p. 10.

few mine managers or supervisors were literate, having learned their trade mainly through apprenticeships, their emergence into the British professional classes was much slower as compared to other occupations. In response, the IMM focussed its efforts on education, and restricted its membership to those with academic credentials. The major difficulty was that the number of members who lived abroad often exceeded those who were residents.[138] The result was an international flavour, but a society which relied heavily on its *Transactions* for communication.

In contrast, the Canadian Mining Institute's objectives were much broader, and it included men from academia and government as well as the mining industry. Membership did not necessarily imply academic credentials, the only difference being if a man was engaged professionally, he was considered a Full Member, and if employed "commercially", he was an Associate.[139] Whether or not one could join, ultimately depended on a system of peer review. Names of prospective candidates were periodically printed in the *Bulletin*, and members were asked to scrutinize the list. The issue of Branch affiliation would continue to trouble the Institute. It was implied, but never made explicit, that Branch members should also belong to the national organization.

The Australasian Institute of Mining and Metallurgy (AusIMM), organized in 1893, had also benefited from the AIME experience, and would come the closest to the CIM in its recruitment policy: "We hope to attract men having practical knowledge of many subjects connected with mining and metallurgy."[140] Even its early origins were similar in that the AusIMM had been formed to protest unjust labour practices in the mining industry, and like its Canadian counterpart, soon refused to become a lobby group. But the AusIMM membership was unstable and growth took longer than in other institutions, even though it, too, had Branches by 1909. Because there was little trade in metals and minerals between Australia and the United States, the AusIMM developed stronger ties with other institutions in the British Empire rather than the AIME. Like the CIM, AusIMM would participate in the Empire Mining and Metallurgical Congresses.[141]

---

[138] Ibid., 111.

[139] Harold Mortimer-Lamb, "Some notes on the History and Recent Development of the Canadian Mining Institute," *JCMI*, vol. XIII (1910): 592-594.

[140] John M. Dew, *Mining People–A Century. Highlights of the First Hundred Years of the AusIMM 1893-1993* (Victoria, Australia: AusIMM, 1993).

[141] The South African Institute of Mining and Metallurgy (SAIMM) was also slow to develop. Established in 1894, no meetings could be held between October 1899 and May 1902 due to the Anglo-Boer War. "SAIMM, 1894-1994," *Journal of the South African Institute of Mining and Metallurgy*, vol. 94, no. 7 (July 1994).

Branches and the Institute

When all was said and done, the Canadian Mining Institute had designed a unique set of By-Laws. By encouraging the formation of local Branches, and later the technical sections and Divisions, it was able to bring the diverse interests of mining, metallurgy, and petroleum people together in one Institution. Each Branch would have a Chairman and a Secretary, as well as any additional officers who might be needed to help carry on its activities. Branches were free to decide their own agenda, but the Institute Secretary had to be kept informed of their activities, or of any action that could compromise the Institute, such as the unilateral lobbying of governments. This would act as a brake in case any Branch got too involved in political affairs. Each Branch also had to adhere to the Charter and By-Laws of the Institute, and the Institute could dissolve any Branch at any time.

Finances figured prominently in discussions concerning Branches during the first decade of the twentieth century. Council at first agreed that seventy-five percent of the dues paid by Branch members would be reimbursed to Branches that needed money to defray "ordinary" expenses. However, Stevenson-Brown, the soon-to-retire Treasurer, felt that until he could report "a good round balance in hand," such requests were impossible. Council also felt strongly that it would be "unwise to adopt a policy that would cripple the parent association… and lead to its bankruptcy."[142] When Willet Miller became president, the "rule" was revised: all Branches would have to be self-sustaining, but a seventy-five percent reimbursement would continue to be available to the Western Branch whose membership was widely scattered. To make up for any revenue short-fall, the Secretary was asked to give more time to the Branches, and promote a wider interest in the national Institute. Unfortunately, this did not solve the problems at the Cobalt Branch, and Cole was asked to approach one of the mining companies for financial support. In the meantime, a committee was nominated to apply for an annual Dominion grant of $5,000 to assist in the Institute's expansion.

Other Institute activities focussed on the revision of provincial mining laws and the education of members as well as students. A committee was formed to study and recommend changes to Ontario's mining laws, under review by an Ontario Royal Commission, in hopes that the mining laws of both Ontario and Quebec could be harmonized. In education, the Institute promoted the formation of Branch libraries in several important

[142] "Proceedings of Fifteenth Annual Meeting," *Transactions*, vol. XVI (1913): 4.

centres. Although it could not make any financial contributions to the academic sector, members were encouraged to assist on an individual basis. When a letter reached the Secretary's desk from the Director of the Kingston School of Mining, Council praised the efforts made by several Institute members and Ontario civil servants to have a new building constructed for the Departments of Chemistry, Mining, and Metallurgy.[143]

Inauguration of the Bulletin

The first issue of the *Bulletin* appeared at the Ninth Annual General Meeting in Ottawa in 1908, and this publication brought the Institute's finances into the black within the fiscal year. Council had cancelled all former ties with independent journals and announced that the new quarterly would be its sole official organ. Edited by the Secretary, Mortimer-Lamb, it would pre-publish papers for discussion before recommending them to the *Proceedings*, and print other matters of general interest to the membership, such as Branch reports and biographical sketches. Members were urged to make free use of its columns for the exchange of information, as long as the contributions furthered the work of the Institute. Reviewed papers were still published in the *JCMI*, and the *Transactions* for 1908 comprised over 700 pages, double the number in 1904. These publications were of such high quality that they were sought by many other organizations.[144]

The *Bulletin* was so successful that by April 25, 1914, Charles Fergie, Chairman of the Finance Committee, reported that the Institute's finances were sound enough to up-grade the quarterly to a monthly. However, this placed an enormous burden on the Institute's resources. Most of its revenue was still generated from dues, and starting in May, an application form was included in each issue to attract more members. To help defray the publication cost, Mortimer-Lamb had secured some advertising revenue, and the post office had agreed that the *Bulletin* could be mailed at the privileged rate of ¼¢ per pound, the going rate for newspapers. The practice of double-printing papers was discontinued, and members were reminded to preserve their copies of the *Bulletin*, as from now on, they would contain the Institute's proceedings. The question of how many papers would be accepted in the *Bul-*

---

[143] "Tenth Annual Meeting," *Bulletin*, no. 2 (May 1908): 9.

[144] The *Journal of the Canadian Mining Institute* (JCMI), the *Proceedings*, and the *Transactions of the Canadian Mining Institute* (*Transactions*) had all been published since 1898. The *Journal of the GMAPQ* was dissolved in 1898, but the *Journal of the Mining Society of Nova Scotia* continued until 1917.

letin from individual Branches, and later, from the technical Divisions, would keep the Publishing Committee occupied.

Prelude to War

In his opening address at the fifteenth Annual General Meeting in Ottawa, in 1913, President Alfred E. Barlow announced that the Institute's total membership had risen to 1,052, of which 754 were full members. In a much quoted passage, he commented that:

> The Canadian Mining Institute represents an industry, not a profession… [and with a] more open membership qualification, there is evidently necessary a much greater degree of that *esprit de corps* which contributes so largely to the success even of those exclusive professional associations.[145]

Barlow elaborated on the national importance of mining compared to other industries that depended directly on natural resources. "The measure of a nation's civilization and progress," he said, "is directly porportional to the development of its mineral resources, especially coal and iron."[146] In terms of production, mining was second only to agriculture in economic output, surpassing that of the fisheries and forest industries combined. When these industries were compared by freight tonnage, mining accounted for 36 percent; forestry was second, and agriculture a distant third. Barlow's observations were notable because the strength of Canada's industries, and the loyalty of its people, would soon be under tremendous pressure.

In 1914, Council was planning another year of meetings and other projects. It was very optimistic because, the previous year, the Institute had assisted with the 12th International Geological Congress which had attracted geologists from more than thirty countries. Discussions were also underway with the Secretary of AIME concerning an arrangement for the simultaneous joint publication of papers. The rationale was to give wider exposure to papers of exceptional technical merit, and the move was facilitated through the special relationship which had developed between the two institutes. While Americans often submitted papers to CIM, Canadians used CIM almost exclusively. It was agreed that both organizations would

---

[145] "Proceedings of Fifteenth Annual Meeting," *Transactions*, vol. XVI (1913): 5.

[146] Ibid., 7.

benefit, and the arrangement should stimulate more contributions by members seeking the widest dissemination and best feedback possible.

At the suggestion of the AusIMM, a Pan-Canadian Mining Meeting was also planned for Canada the following year. The meeting was intended to be "thoroughly British," with as many delegates as possible from Australasia, the Malay States, India, South Africa, and Great Britain. The Secretary of AusIMM wrote that even though it was difficult to arrange passage from halfway around the globe, several of its leading mining men were bent on attending. Aware of the growing unrest in Europe, he added that "Such a gathering of mining men from all parts of the Empire, in Canada, should be not only of great service from a mining standpoint, but fulfil a political purpose as well."[147] Although well-intentioned, the meeting would soon be cancelled.

Barlow had agreed to bring the Pan-Canadian idea before the IMM in May, as he and his wife were anticipating a long-awaited trip to England. Tragically, their journey would never be completed. The Barlows had boarded the ill-fated *Empress of Ireland*, a Canadian Pacific ocean liner which sank in the St. Lawrence near Rimouski on May 29, 1914. Rammed in dense fog by the Norwegian collier *Storstad*, it sank within fourteen minutes. Only 465 people survived, leaving 1,014 passengers and crew to be buried in a mass grave at Métis-Sur-Mer, Quebec. When news reached the Institute that the Barlows were among the deceased, letters of shock and sympathy poured into the Secretary's office. The couple had always been admired and well-liked, leading one writer to comment: "He went down with his helpmate and comrade, probably fighting for their two lives. It is somewhat a consolation to their friends to feel that neither of the two would have cared to survive the other."[148]

Barlow's reputation as one of Canada's top geologists in the Geological Survey was legendary, he having predicted Cobalt as a "promising mining field" in the late 1890s. He was also an authority on Sudbury's nickel and copper deposits, as well as the Archaen rocks of the Haliburton and Bancroft districts. He was the first of the Institute's Past Presidents to be mourned, and shortly after his death, an award was endowed in his memory. Of his special relationship to CIM, Mortimer-Lamb wrote:

> Not only during his term as President, but for many years it was his custom to visit headquarters almost daily and to

[147] J. Mackintosh Bell, Past President (1911) of AusIMM, letter to the CIM Secretary, September 18, 1913, vol. 3, NAC.

[148] Theo Denis, letter to the Secretary, June 9, 1914, *Bulletin*, no. 27 (July 1914): 5.

> spend his time ungrudgingly in assisting in such tasks as the editing of material for publication or in discussing with the writer questions of office detail... The Institute owes its existence and its continued usefulness to [his] whole-souled, self-sacrificing zeal. The loss to the Institute is very great.[149]

Indeed, a succession of deaths would affect the Institute's members in much the same way as their disbelief over the news of a world war. During the previous few months, members had mourned several prominent people, among them Fritz Cirkel, Archibald Blue, and W. F. Sutton, a distinguished geologist, Institute Vice President, and Chairman of the Western Branch. Also lost was Robert G. Leckie, whom Mortimer-Lamb had fondly described as "the father of mining engineering in Canada."[150] With Barlow now gone, Council could not bear the thought of losing another good man. On September 21st it received a letter of resignation from John J. Penhale explaining that he had enlisted in the Divisional Ammunition Column, and as its Commanding Officer, would soon be headed to the Continent. Penhale was not given to sentimentality, but his words were foreboding:

> The CMI does not need 'dead ones,' neither actual or potential... If it turns out otherwise, I hope you will remember that even if I have not accomplished much for the Institute, I have always done my best to help out. The friendships formed at the various meetings have been, and are bright spots in my memory, and I hope they will not be blotted out too quickly.[151]

Under the circumstances, Council unanimously decided *not* to accept his resignation.

The First World War

On August 4, 1914, Canadians awoke to the startling news that Britain had declared war against Germany. This information came in a rapid

---

[149] Mortimer-Lamb, "The Late Dr. Alfred E. Barlow," *Bulletin*, no. 27 (July 1914): 3-13.
[150] "The Late Major R.G. Leckie," *Bulletin*, no. 23 (January 1914): 17-19.
[151] John J. Penhale, Valcartier Camp, Quebec, letter to the Secretary, September 21, 1914, vol. 3, NAC.

succession of press releases which detailed the assassination of Archduke Ferdinand in Sarajevo, the ambitions of Kaiser Wilhelm II, and the uneasy borders of the Balkan kingdoms and Russia. Although Canadians knew of the growing tensions in Europe, many believed that they were insulated by the distance. Safely located an ocean away, they had been enjoying steady economic progress since 1911, and were generally confident about their future. For many, it seemed incredible that within hours of the declaration, their lives would be forever changed.

As a member of the British Empire, Canada's entry into the war was automatic. Within three weeks, Ottawa alone was besieged by more than 100,000 young men who had eagerly volunteered. Many thought that the war would be a great adventure of short duration. On October 3rd, 32,665 men and 8,500 horses of the Canadian Expeditionary Forces boarded ships for England.

The Princess Patricia's Canadian Light Infantry was the first to see action, taking their positions in the craters of St. Eloi on January 4, 1915. Three months later, the balance of the Canadian Expeditionary Force fought as a division at the Ypres salient in Belgium. Nothing could have prepared these men for the grim existence of trench warfare and the heavy artillery fire and poisonous gas that resulted in thousands of casualties. A second Canadian contingent arrived in France in September, followed by a third in December, and by spring they engaged in the Battle of Mont Sorrel. Next came the lengthy Battle of the Somme where the inhumane process of capturing worthless holes in the ground left a total of 24,029 men dead. By April 1917, the action focused on the Battle of Vimy Ridge, which cost 11,000 Canadian lives in a span of five days. Another 9,000 were lost the following year when the Battle of Amiens, on August 8, 1918, finally witnessed the turning point of the war.

Starting in 1914, the Institute had passed a resolution that membership dues would be waived for all those serving with the Colours. As this eventually affected about twelve percent of the Institute's membership, it was hoped to offset the reduced revenues by asking civilian members to pay their dues promptly, and to encourage new men to apply. In return, the Institute vowed to continue with business as usual, which meant supporting Canada's mining industry through its technical programs, and bolstering the morale of those most affected by the war. According to Mortimer-Lamb, the mining industry would undergo hard times because "the metalliferous mining in all branches, save that of gold, will feel for a period not yet determinable, the [war's] injurious effects..."[152] His prediction was true initially, but by 1917 the unrelenting demand for 'war' metals and higher productivity had pushed a reduced technical and labour force to the breaking point. As profits soared in nearly every mining sector, the stress of increased production caused instability in the marketplace, and industrial workers, discontent with wages and conditions, staged their protest in rotating strikes. These labour problems would persist into the next decade.

The Tunnelling Companies

The number of men affected by Mortimer-Lamb's "hard times" prompted the Institute to raise a corps of mining engineers for active service. Following the lead charted by the IMM in 1914, George G. S. Lindsey, as President, urged all those interested to contact the CIM offices. By 1915, the First and Second Canadian Tunnelling Companies had been recruited in Canada, and a Third was organized in France. All three Companies were officered by CIM members. Some Canadians who were already overseas joined with other mining engineers from the IMM and the AusIMM to form the British Tunnelling Companies. Their strength was numbered in 700 officers and 20,000 men in the non-commissioned ranks. Together, they adopted a much more aggressive method for advance, mining under the German strongholds to detonate massive charges of explosives. In 1917, the Canadian tunnellers, who took part in the capture of the Messines-Wytschaete ridge near the Ypres salient, were praised by their Commander-in-Chief for "the greatest offensive mining operation in the history of war."[153]

Letters from the Front

The real test of a professional organization must surely be its behaviour during a prolonged crisis. Evidence of the Institute's commitment to carry on its mission despite the war can be found in the pages of the *Bulletin*, which averaged more than one hundred pages per issue during this period. Not only was it a medium for the exchange of technical information and ideas, but a venue for members serving overseas to write about the battles, express their feelings in poetry, and share their grief for lost or missing friends. From "somewhere in Belgium," Col. John J. Penhale wrote:

[152] H. Mortimer-Lamb, "The War," *Bulletin*, no. 29 (September 1914): 3-5.
[153] A.W. Davis, "Mining in Warfare," *Bulletin*, no. 172 (August 1926): 907-920.

> I have been hoping that each Canadian mail would bring me some news of the Annual Meeting of the Canadian Mining Institute but no such luck. Funny, isn't it, that one should think of such a thing at such a time, but I wondered where you had decided to have the next meeting; and tried to imagine what each of you were saying and doing... In fact I must confess to a sort of homesick feeling that week.[154]

The letter reported that he, the Leckies, "Billy" King and other good friends came through the "Hell of Ypres and St. Julien" unscathed, and that he would enjoy receiving the *Bulletin* in care of the British Expeditionary Force, 1st Canadian Division. Towards the end of the war, another member would comment that "The *Bulletin* is the best source of Canadian news I get."[155]

Mortimer-Lamb personally acknowledged all the letters from the Front, which helped soothe the isolation and hardship felt by those on active duty. Members at home were inspired by reading of the exploits of their colleagues. A special section, called "C.M.I. Members at the Front," began to appear in the *Bulletin* in July 1915 and became a regular feature. Excerpts from the letters provide a slice-of-life that few people living in recent times can fully appreciate:

> We are out of the trenches for four days in Divisional reserve... It's raining Billie-be-damned now and the men in the trenches have to wear long hip boots and jerkins... We have lots of work to do to keeping the trenches in shape, as so many cave in with this wet weather... I heard a Canadian battalion marching past tonight in the rain and mud singing like the good 'uns.[156]

While the receipt of tobacco and cigarettes from the Tobacco Fund established by the Institute provided some comfort, it also became a means of ascertaining whether certain men were still alive or not. Their brief notes of gratitude at having been remembered by the Institute were printed in

[154] John J. Penhale, "Somewhere in Belgium, 2 May 1915," *Bulletin*, no. 38 (June 1915): 365-66.
[155] Lieut. K.D. Woodworth, *Bulletin*, no. 75 (July 1918): 637-638. The "Leckies" refers to the two sons of R.G. Leckie: Lieut.-Col. R.G.E. (Bert) Leckie and Captain J.E. Leckie, DSO. "Billie" King was Brig.-Gen. W.B. King, DSO, a well-known member from the Cobalt Branch.
[156] Col. J.E. Leckie, in France, in the *Bulletin*, no. 45 (January 1916): 56.

the *Bulletin* for all to read. In fact, some members missed the Institute so much that they organized their own 'annual' dinner:

> As you know there is a big crowd of the 'boys' over here, and I think we can round up most of them. I wish you could drop in for the evening. It won't be as comfortable as the Ritz, but, nevertheless, we hope to make things hum.[157]

The Flanders Branch

In May, a humorous exchange appeared in the *Bulletin.* Col. J.J. Penhale announced that a CIM Branch had been formed in Flanders. Mortimer-Lamb replied that according to Section 54 of the By-laws, no Branch could be formed without the consent of Council, and he was to return home immediately to be "disciplined." Penhale replied:

> I will admit at once that the request... appealed very strongly to me. I would gladly submit to any inquisition I could possibly be subjected to, if it came at the end of a journey *Home*... I'm afraid you will have to accept the Flanders Branch as it stands.[158]

He then outlined the evening's proceedings of the "First (and we hope last) Annual Dinner in the Dug Out, Thursday, March 2nd, 1916, somewhere in France," which included an excellent dinner, "a rattling good speech" by Major-General Arthur W. Currie, Commander of the First Division, and a rendition of "Drill Ye Terriers" by R.W. Brock, whose "ideals were high and so was the 'pitch' in which he led us in singing the C.M.I. anthem." He added that they were "keenly disappointed because quite a number who intended coming were prevented from doing so at the last minute... [but] many incidents and stories of the Sessions at home were told and for a few hours we lived, happily, in the past."[159]

Throughout some of the heaviest military offensives, members wrote of their attachment and longing for the Institute: "It is very gratifying to feel that we, over here, are not out of touch with our confreres who are

[157] Capt. G.C. Riley, *Bulletin*, no. 48 (April 1916): 371.

[158] "The Flanders Branch of the Institute," *Bulletin*, no. 49 (May 1916): 431-439. The 'unofficial' Flanders Branch had been in existence since August 1914.

[159] Ibid.

still in Canada..."[160] At the time, many believed they would be home for St. Patrick's Day:

> The Institute means a lot more to me than it did twelve months ago; and I think most members out here feel the same way about it... Up to the present I have been rather fortunate in escaping damage. I enjoy the best of health and keep my head well down... I am looking forward to being present at the 'big spread' in 1918.[161]

Other letters described in graphic detail the now famous campaigns at Vimy Ridge and Messines, in the spring of 1917:

> I have had a trying winter on Vimy Ridge, but, in general, I have been exceedingly fortunate, notwithstanding, mud, cold, lice, rats and 'rum jars' - I name the horrors in order of merit... The Canadian victory at Vimy has created a universal impression. To those of us, who have reason to know that locality fairly well, the victory is all the more remarkable.[162]
>
> Should I be lucky enough to come through safely I shall never regret the opportunity I have had of working with our miners... the Canadians who had mined Hill 60 [at Messines] had every reason to be proud of their work... the Germans in the front lines have been given such a shaking that they imagine mines everywhere...[163]

The correspondence did not cease when the war ended. Members detained in Europe during the next two years would continue to express their appreciation to the Institute. Through the pages of the *Bulletin*, the Institute had done more for its membership then it had ever believed possible.

> In particular I express gratitude for the way in which the C.M.I. has looked after our comfort and kept us in touch with one another through the *Bulletin*... I also express appreciation of the fact that the profession has backed up

[160] Capt. A.G. Macaulay, *Bulletin*, no. 55 (November 1916): 1005.
[161] Sergeant W.F. Gowans, No. 1 Tunnelling Co., *Bulletin*, no. 56 (December 1916): 1068-1070.
[162] Major J. Mackintosh Bell, *Bulletin*, no. 62 (June 1917): 523.
[163] Lieut. G. Lewis Burland, *Bulletin*, no. 64 (August 1917): 679-680.

> the Canadian Army in every possible way; and that the mining engineer has found his civic soul in the national life of Canada… Permit me to wish that the *après la guerre* meeting of the C.M.I. will be the most successful in history.[164]

End of the Great War

After the decisive Battle at Amiens, in August 1918, the Allied efforts to defeat the enemy gained new momentum. Within three months, Kaiser Wilhelm II had abdicated his throne, the German government had surrendered, and the long-awaited Armistice was signed on November 11, 1918. The devastating cost in human lives on both sides was estimated at over ten million, with enormous numbers wounded or permanently disabled. Of the 418,052 Canadians who had gallantly served their country, 60,000 were killed, 218,600 were wounded, and another 30,000 were still recuperating in British, French, or Canadian hospitals.[165] The toll among the civilian population will never be truly known.

The Honour and Service Role

The Institute's Honour and Service Role, inaugurated at the Annual Meeting in March 1919, listed 125 members in active service, eleven of whom made the ultimate sacrifice. Consisting of an oak cabinet nine feet tall, its unique design allowed the brass plaques to be swung open to reveal the precious photographs underneath. It stood for many years in the Institute's Headquarters, which had moved in the fall of 1917 to the Drummond Building in downtown Montreal.

Among the officers and soldiers of the combined tunnelling units who did not survive were CIM members from the Kootenays, Vancouver Island, northern Ontario, and Cape Breton. The fortunate included Allen A. Mackay, George E. Cole, and Alan E. Cameron, all future CIM Presidents, along with James G. Ross, a Lieutenant with the 13th Battalion, Royal Highlanders, who was wounded at Festubert. The Princess Patricia's Canadian Light Infantry suffered particularly heavy losses. Only forty-four survived out of the 1,098 who had enlisted in 1914. Of these, James B. Mawdsley was wounded twice and was

[164] Capt. I.D.D. Daimpre, L'Havre, France, *Bulletin*, no. 83 (March 1919): 320.
[165] *Bulletin*, no. 86 (June 1919): 636-639.

awarded the Member of the British Empire (MBE). He would become CIM President in 1961-62.

Any loss of life is tragic, but the death of Captain Osmond E. Leroy at Passchendaele, Flanders, on October 26, 1917, was particularly hard to accept. Born in Argenteuil, Quebec, in 1873, Leroy had graduated from McGill and was rapidly becoming one of the foremost economic geologists of his time. A well-known member of CIM, he had trained under both Frank D. Adams and Alfred E. Barlow, and had spent three years in China as a consulting geologist. With a brilliant career ahead, his sense of humour and natural ability had kept many members entertained during the annual meetings and 'after-meetings.' The Institute was shaken by his death, as he represented its very best and brightest. Leroy's five-page obituary in the *Bulletin* concludes with this epitaph:

> His life was gentle, and the elements
> so mixed up in him that Nature might stand up
> And say to all the world 'This was a man.'[166]

The Khaki University

In the spring of 1919, Canadian troops awaiting demobilization in Britain were able to enrol in the Khaki University. Many could resume their interrupted university studies, and others could pursue a high-school matriculation. Known as the Educational Service of the Canadian Military Forces Overseas, this correspondence school was directed by Henry M. Tory, President of the University of Alberta, and Frank D. Adams, Dean of the Faculty of Applied Science at McGill University and a CIM Past President. Taught by specialists who were readily available, such as Alan E. Cameron, a professor of mining engineering, the school offered courses ranging from literacy to various areas of science and engineering. This initiative, on the part of the Canadian government, helped to atone for lost time. Among the beneficiaries of this program was Albert V. Corlett, who served in the Royal Canadian Engineers and was able to complete his second year of university before returning home. Corlett subsequently became an active member of CIM, and was Head of the Mining Engineering Department at Queen's University from 1948 to 1965.

[166] S.J. Schofield, "The Late Captain Osmond Edgar Leroy," *Bulletin*, no. 70 (February 1918): 149-155.

A Royal Institute

There is little doubt that the Institute had played an extremely important role during wartime. As a symbol of its patriotism, and to show support for its servicemen overseas, Council had resolved, in 1916, to ask Parliament for a Royal Charter which would have renamed the Institute the "Royal Canadian Mining Institute." Although the proposal was printed in the *Bulletin*, it did not generate much interest among the membership. When the war in Europe intensified, the timing was said to be inappropriate and the idea was later dropped. Perhaps the Council thought that actions were better than words, and when compared to the courageous efforts of its members, the word "Royal" would be superfluous.

Après la Guerre

The Annual Meeting held in Montreal in 1919 was a time to reflect on new needs in the areas of employment and training. One was the placement of returned soldiers, many of whom were impaired. Council urged that great care be taken to give veterans a fair, if not preferential opportunity for jobs, and employers were asked to loosen their job requirements. For example, of the 3,000 miners who had enlisted from Nova Scotia, 4.5 percent were expected to return totally deaf. Such men were still capable of operating mining equipment underground with few risks to themselves or others.

For men who wished to further their education, new opportunities were opening up. Before the First World War, the formal structure of Canada's science and technology was limited, even though some scientists, like Frank D. Adams, had invested their time in projects such as the Canadian Commission of Conservation (1909-1921).[167] When a federal order-in-council launched the Honorary Advisory Council for Scientific and Industrial Research in 1916, later known as the National Research Council (NRC), the Institute was clearly in favour, having already anticipated the needs of industry some years before.[168] Two ideas that would become the hallmark of the NRC were a scholarship and grant program to assist individuals and

[167] Michel F. Girard, "The Commission of Conservation as a Forerunner to the National Research Council 1909-1921," *Scientia Canadensis*, vol. 15, no. 2 (1991): 19-40.

[168] Richard A. Jarrell and Yves Gingras, Building Canadian Science: The Role of the National Research Council, *Scientia Canadensis*, vol. 15, no. 2 (1991): 1-17. And Frank D, Adams, "The Work of the Advisory Council for Scientific and Industrial Research in Canada," *Bulletin*, no. 61 (May 1917): 407-415.

universities, and the notion that specific problems related to industry could be tackled using the scientific method.

Educational Reform

Concurrently, C.V. Corless , a humanitarian, and the General Manager of the Mond Nickel Company, had begun to scrutinize Canada's educational system and its connection with unrest between capital and labour. His series of articles on educational reform, published in the *Bulletin*, were touted as "the most important ever presented to the Institute." Although their subject was not technical, the Institute believed that these papers would "reach deeper down and wider outwards than any technical discussions we could have."[169] Corless, who would become a CIM president in 1921-22, felt strongly that every Canadian student needed instruction in the social sciences:

> To develop in him that liberal and liberating view of all honest work and effort, which adds to it the dignity and pleasure of social service and removes from it the feeling of servility and reluctance that arises from having no other interest in one's occupation than the wage or salary or other material reward.[170]

The excitement generated by Corless' initiative resulted in the drafting of a new Committee on Education in 1918, composed of representatives from the CIM, The Canadian Pulp and Paper Association, McGill University, Macdonald College, the Engineering Institute of Canada, the Canadian Manufacturer's Association, the Trade and Labour Council, the Society of Chemical Industry, and the Canadian Fisheries Association. After several meetings, the Committee presented a comprehensive resolution to the Minister of Education which recommended everything from citizenship training, to an increase in teachers' wages, better text books, the establishment of school libraries, compulsory education to age fourteen, and the "systematic and scientific planning of settlements of unoccupied lands,"

[169] C.V. Corless, "Educational Reform - Its Relation to a Solution of the Industrial Deadlock," *Bulletin*, no. 68 (December 1917): 976-984. C.V. Corless, "Training for Social Responsibility," *Bulletin*, no. 61 (May 1917): 401-407.

[170] C.V. Corless, "Educational Reform, Part I," *Bulletin*, no. 69 (January 1918): 43-64. Also Part II, "Educational Reform - Its Relation to a Solution of the Industrial Deadlock," *Bulletin*, no. 70 (February 1918): 128-140, and *Bulletin*, no. 71 (March 1918): 210-224.

which would result in more centrally-based communities with larger, better equipped schools. With its involvement in this committee, the Institute's concern and support over education developed into a lasting interest.

The Mining Society of Nova Scotia Rejoins the Institute

The cordial relationship between the Institute and the Mining Society of Nova Scotia had steadily improved, particularly because the majority of mining men in the Maritimes were also CIM members. The feasibility of a common headquarters and library had already generated some interest, and in 1908, the Institute had approached the Society offering to publish its *Transactions*. While the amalgamation of the two organizations was still not a consideration at that time, several members entertained the idea of a common reading room, where books and periodicals would be accessible to everyone. Not everyone shared this viewpoint. Some members of the MSNS, such as Robert Drummond, continued to remind the Society of older issues: "There are certain things this Society is not prepared to do. It is not going to become a branch of the Canadian Mining Institute, to lose its identity and become swallowed up in that body..."[171]

At the MSNS's Annual Meeting in March 1912, CIM President Alfred Barlow had raised the issue of affiliation, which was immediately taken up by Frederick H. Sexton, the Society's incoming President and a loyal Institute member. He agreed that the time was opportune to bring the issue to a resolution, and the Society's Council was empowered to enter into discussions with the Institute. The process would be a long one.

In April 1914, CIM President George G.S. Lindsey submitted the Institute's revised proposal for affiliation to the MSNS. A lawyer by profession, he had become President of the Crow's Nest Pass Coal Company, and had spent part of his time in China helping to draft a mining act for the new Republic. The fact that he was a skilled negotiator, and a coal man, would benefit the Institute's relationship with the MSNS. The proposal stipulated that the MSNS would have the right to maintain its own name, the right to self-government, and the right to retain any grants or donations given to the Society. On its part, CIM would accept every member recommended by the MSNS, and in return for the $10 membership dues, it would assume the cost of publishing the Society's papers and reports in the *Bulletin* and *Transactions*. Before serious negotiations could proceed, the MSNS

[171] Pigot, 65.

announced that because of the war in Europe, it would suspend its annual meetings, which essentially removed it from the bargaining table.

It was not until 1917 that discussions were resumed with the appointment of a MSNS committee consisting of Thomas J. Brown and Francis W. Gray, both future CIM presidents, George D. McDougall, a future president of the MSNS, and Sexton as chairman. The terms of affiliation were very similar to those originally proposed by Lindsey in 1914, and in July 1918, the *Bulletin* reported that only a few minor details were left to negotiate. Of these, the major point was a $250 maintenance grant from the Institute, which the MSNS felt was essential to replace lost revenue from membership dues. The Institute stalled for a while, hoping that the figure would drop, but then finally conceded. At a CIM Council meeting on November 15, 1918, the terms were accepted by both parties, and the affiliation became an established fact.[172] It had taken twenty years for the Mining Society of Nova Scotia to join with the Canadian Mining Institute.

A Coming of Age

For the first time in its history, the Institute had become a truly national organization, and could claim to represent all of Canada's mining, metallurgical, and petroleum professional interests from Cape Breton to Vancouver Island. The Institute had also expanded in other ways. A Metallurgical Section had been organized on April 7, 1916, an Iron and Steel Section on April 5, 1918, and a Coal Section, realized in 1920, was in the planning stage. In addition, three new Branches had been established in Manitoba, New Brunswick, and Vancouver, and another in Sudbury awaited Council approval. Institute membership had risen to more than 1,000, and an assistant had been hired to help Mortimer-Lamb with secretarial duties. It would not be long before members of the Institute and of the MSNS would press the Institute to adopt a more inclusive name.

In his presidential address in 1919, Donaldson B. Dowling referred to these changes as "the coming of age of the Institute."[173] Like Canada and its new found identity, the Institute had shed its youth and was embarking on a new age of responsibility and leadership. Many men who had entered the Great War as British subjects returned as Canadians who had fought shoulder-to-shoulder against adversity. For those who had operated the

[172] Pigot, 65-70.
[173] "Proceedings of Twenty-first Annual Meeting," *Bulletin*, no. 84 (April 1919): 351.

mines, mills, and factories on the home front, a sense of national pride helped ease, for a time, the effects of rising labour disputes and inflation. For Dowling, the most pressing issues were technical education, job training, and the wise use of natural resources. Having survived and flourished during its first twenty years, the Institute now needed to consider its future. In his progress report for 1919, Mortimer-Lamb wrote:

> If the forefathers of the C.M.I. dreamed dreams and saw visions, they set us a good example, because in Canada there lies the substantial stuff from which vision can be translated into actuality [by] this generation of Canadians… No technical society can look forward to its future with such justified optimism as should guide the policies of the Canadian Mining Institute, and inspire the enthusiasm of its membership.[174]

For the Annual Meeting in 1914, an official lapel pin had been produced bearing the Institute's crest in blue enamel and silver. Members were now encouraged to wear it with pride, as belonging meant much more than mere distinction ever could.

The Canadian Institute of Mining and Metallurgy

Although metallurgy and metallurgists had been an integral part of the Institute from the very beginning, the growth of a distinct metallurgical industry in Canada, the affiliation of the MSNS, and the Institute's new Iron and Steel Section, and Metallurgical Section, all provided impetus for a name change. An Institute more inclusive in name, if not in fact, would also attract new members. The matter was brought to a head by Alfred Stansfield. At the Annual General Meeting, Stansfield introduced a motion that the Institute's name be changed to the Canadian Institute of Mining and Metallurgy. He noted that "as time has gone on Canadian metallurgy has become more diverse, more important, and less closely related to mining." George C. Mackenzie, who was shortly to become the Institute's Secretary-Treasurer, seconded the motion:

> If this Institute is to keep abreast of the times, it would be a wise measure to broaden its title. From my own experi-

[174] *Bulletin*, no. 91 (November 1919): 1114-1115.

> ence in talking to my friends in the metallurgical industry, suggesting that they should be members of this Institute, I am met with the statement that the Institute is purely a mining body and... [that] consequently, they believe that very little would be gained through membership.[175]

A lively discussion ensued. E.J. Carlyle, H. Mortimer-Lamb, F.W. Gray, and others spoke in favour of the name change. Even though an Iron and Steel Section had recently been organized within the Institute, Mortimer-Lamb noted that "some iron and steel men refused to join this section on the ground that they had no direct connection with mining," and Gray pointed out that "if we do not make the [name] change, the iron and steel men may... form an institute of their own" as they had in England. Mackenzie added that "in view of the fact that metallurgists in Canada have made such advances as they have in recent years, I think some recognition should be given them by the Institute."

Among those opposing the motion were such prominent members as Willet Miller, John A. Dresser, and H.E. Haultain. Miller argued that the "term mining includes metallurgy, as well as geology... [and he] did not see how the proposed change is going to encourage the metallurgical industry." Haultain felt that "certain traditions had grown around [the Institute's] name ... and these would be shattered if there were any change."

A name change was clearly contentious, and it was proposed by an amendment that the matter be put to the Institute's entire membership by a letter ballot. The vote on the amended proposal was close: thirty-six in favour, and thirty-two against. The result of the letter ballot was much more favourable: 343 in favour of the name change, and 92 against. The addition of the word "Metallurgy" and the abbreviation C.I.M.M. satisfied most members. The Institute's new name was formalized by an Act of Parliament on May 11, 1920. Seventy years later, the Institute's members would again be asked to approve a name change.

[175] "The Proposal to Change the Name of the Institute," *Bulletin*, no. 85 (May 1919): 506- 513.

THE MIDDLE GROUND: 1920-1966

Starting in 1920, the Institute took much stronger steps towards the promotion of the mining industry, the welfare of its members, and the dissemination of technical knowledge. Its early relationship as a critic of government had gradually matured into one of confidential advisor, largely through the initiation of a federal Department of Mines.[176] Issues of status and policy would continue to be hotly debated, but there was a prevailing sense of change in the Institute. Many of those who had founded the Institute and directed its progress were quietly disappearing. William H. Merritt had succumbed during the flu epidemic of 1918. George G.S. Lindsey, the architect of affiliation between the Mining Society of Nova Scotia and the Institute, died in 1920. George R. Smith would pass away in 1922. John Hardman would write the obituary of Charles Fergie in October 1924, shortly before Willet Miller's death in February 1925, and his own in April. Donaldson B. Dowling, of the Geological Survey, would follow them in July.

People with different backgrounds and ideas were coming forward, and Council began to envision the Institute as a powerful mining industry organization. The Institute's name change in 1920 would create new expectations from the membership, and new specialized technical groups, designed to function much like branches, would shortly be established. A Metallurgical Section had already been formed in 1916, an Iron and Steel Branch in 1918, and a Coal Section in 1919. The Petroleum Section, appearing in 1926, would be the best organized of the four. An Industrial Minerals Section, the first to be created under official new guidelines, would follow in 1932. The Branches would begin to be organized into geographic divisions, now known as Districts. The first of these had already been formed in 1919 as the British Columbia Division, although it would not be officially recognized by the Institute until much later.

A New Secretary-Treasurer

In 1919, Mortimer-Lamb resigned as Secretary-Treasurer due to ill health and exhaustion. He had served the Institute well for seventeen years, and many executive members had regarded him as their "hardy perennial." Within the year, a more restored Mortimer-Lamb, now Secretary-Emeritus

[176] Mortimer-Lamb, "The Institute in its Relation to Government," *Bulletin*, no. 107 (March 1921): 231-36.

of the Institute, became Secretary-Treasurer of the newly formed British Columbia Division, and would provide his support and expertise for the next twenty-five years. He continued to edit the *Bulletin* and the *Transactions* from the west coast until George Cleghorn Mackenzie succeeded him on January 1, 1921. Mackenzie would hold this post for the next ten years, although Council would reappoint him only on a yearly basis.

Mackenzie was a familiar Institute man, having become a student member in 1902, a full member in 1907, and a councillor in 1916-17. Born in Ontario's Bruce County in 1877, he had a BSc in mining engineering and had carried out geological field work under Alfred E. Barlow. As his interest in metallurgy grew, he worked for the Canada Iron Furnace Company of Midland, Ontario, and after graduation, was appointed assistant blast-furnace superintendent at the Londonderry Iron and Mining Company in Nova Scotia. By 1907, he was on the staff of the Ontario Bureau of Mines as a claims inspector; subsequently, he joined the Mines Branch in Ottawa. When the opening for a secretary-treasurer was advertised in 1919, he was general manager of the Electro Steel and Engineering Company of Welland. Because Mackenzie's background was quite different to that of his predecessors, the Institute dropped the editorial duties from the job description when he was appointed Secretary-Treasurer. Shortly afterwards, Professor Richard P.D. Graham became Technical Editor for Institute publications, a voluntary position he held for the next forty years in addition to his teaching duties at McGill University.[177] Mackenzie's tenure would encompass some of the most challenging years of Institute management.

As Canadians struggled with the transition of a wartime to a peacetime economy, their lives would be drastically altered by new concepts and commercial products. On August 17, 1918, the country's first airmail travelled by biplane from Toronto to Ottawa. On May 20, 1920, the first live radio broadcast of over a hundred miles was transmitted from Ottawa to Montreal, ushering in the age of public radio. Factories which had maintained a steady production of tanks, munitions, and other military supplies, retooled to produce automobiles, tractors, electrical appliances, and civilian aircraft. In government, an array of social programs had been introduced, and in labour, there had been a marked increase in participation in trade unions.

The boom in consumer goods was accompanied by higher costs for food and necessities. Wages had been adequate, but as export markets began

[177] R.P.D. Graham began as Technical Editor in 1921, then was officially re-appointed from 1950 to 1962 after his retirement from McGill University in 1950.

to slide, they could not keep pace with rising inflation. The manufacturing, construction, and transportation industries protested the rise in tariffs on certain products exported to the United States, and unemployment rose as many companies went bankrupt. The scene in 1919 was of rotating strikes among the metal and building trades, of which the Winnipeg General Strike was the most serious, affecting more than 30,000 people for several weeks. Although the recession did not affect everyone equally, farmers, industrial workers, and small-business owners were hardest hit. Some of those searching for work, or better opportunities, migrated to the United States. The Canadian economy did not rebound until after 1924.

The Institute's Employment Bureau

As jobs were harder to find, the placement of skilled war veterans in the workforce required special consideration. Institute members returning from overseas were encouraged to communicate with the Secretary, and forward their resumes to the Institute's offices. The monthly *Bulletin* printed a column of eligible personnel, classified by number and experience, next to job advertisements. Unfortunately, the number of those looking for work vastly outnumbered the possible opportunities. Most mining companies across the country supported the initiative, which proved beneficial. However, disabled veterans needed special assistance.

To his immeasurable credit, Herbert E.T. Haultain launched an extensive program of occupational therapy for disabled veterans in Ontario hospitals, and personally supervised the training of the therapists, most of whom were women. Haultain also founded the Technical Service Council of Ontario, an organization which encouraged engineers to remain in Canada. Haultain had earned his first degree at the University of Toronto in civil engineering before enrolling in postgraduate courses in England and Germany. He had gained a lot of practical experience, having worked at pick-and-shovel jobs, and then managed mining projects around the world. In 1908, he was appointed Professor of Mining Engineering at the University of Toronto, where he would teach for the next thirty years. A longtime member of the Institute, he served as a Vice President in 1918-19. Although Haultain could be "gruff, outspoken, and controversial" in his approach, he was basically a kind man who shunned academic and professional recognition. The only award he would ever accept was the Institute's Randolph Bruce Gold Medal, which recognized his outstanding research in ore dressing. Many Canadian mining and metallurgical engineering graduates still remember having used a Haultain Infrasizer in laboratory courses. Posthumously, Haultain would be inducted into the Canadian Mining Hall of Fame in 1994.

Iron Ring Ceremony

Haultain is best remembered for his creation of the Iron Ring Ceremony. Inspired by the Hippocratic oath in medicine, he searched for a similar ritual to guard the ethics and integrity of the engineering profession. Having been a great admirer and personal friend of Rudyard Kipling, Haultain approached him for assistance in writing a suitable pledge. The result was what Canadian engineers now call the "Ritual of the Calling of an Engineer." As tangible evidence of an engineer's commitment to his vocation, Haultain introduced an iron ring, to be worn on the little finger of the engineer's working hand. For many years, these rings were fashioned by convalescing war veterans, under the guidance of the occupational therapists.[178]

Debate Over Membership

In 1921, Mackenzie embarked on an extensive six-month tour of Branches, and tried to gauge the health of the Institute in terms of membership and finances. In his report to Council, he proposed a series of policy changes which Mortimer-Lamb would soon claim as "radical" and, if adopted, would "divert entirely, for good or ill, the direction in which, during the past number of years, the Institute has been travelling."[179] The fundamental question was whether the Institute was willing to accept all members on an equal footing, and abolish the distinction drawn between the categories of associate and full membership. In a position paper published in the January 1922 *Bulletin*, Mackenzie rehashed the many opinions he had heard concerning the Institute. It was not functioning in accord with the original Charter, and newcomers felt that it was becoming exclusive and class-based. He wrote, "How can this Institute accomplish anything of tangible value to the industry garbed as it is with a coat and trousers of academic cut and a modest waist-coat of professional tone?"[180]

Mackenzie felt that the time had arrived to redefine the Institute's goals and decide, once and for all, on the criteria for membership. At the

[178] L.E. Jones, "Delineations of Destiny, Herbert Edward Terrick Haultain (1869-1961)," *Engineering Digest*, vol. 27 (July 1976): 28. "Canadian Mining in Debt to Prof. H.E.T. Haultain," *Financial Post* [Toronto], October 15, 1932.

[179] H. Mortimer-Lamb, "Institute Policy," *Bulletin*, no. 119 (March 1922): 396-405.

[180] G.C. Mackenzie, "Institute Policy," *Bulletin*, no. 117 (January 1922): 23-33.

time, the CIM was neither a professional nor an industrial society, but a mixture. Since 1907, the By-laws had distinguished between technical or professional, and non-technical men by classifying them as full or associate members. There was little practical difference between the two, except that an associate could not hold office.

In the *Bulletin*, Mackenzie reminded readers that the Institute was organized for the purpose of representing and encouraging the mining industry. When the Charter was written in 1898, there was no mention of vested interest, nor was any distinction made among various engineering specialties. He then pointed out that there were many individuals who were certainly eligible to join the Institute, but who did not make the effort. Of the 1,576 members in 1921, 88% were Canadian residents. Mackenzie compared this number to the 1920 Canadian Census figures, in which 72,888 people listed their occupations as being within the mineral and metallurgical industries. Of the total, about 7% reported having positions in middle or upper management, but only 2% of these actually belonged to the Institute. Mackenzie also pointed out that the statistics did not include:

> public servants, University men, professional men in private practice, nor the much larger number of non-professional people who have financial investments in these industries – all of which classes are represented in the Institute membership.[181]

Most people understood that the Institute needed a modest bank account to achieve moderate growth. If membership was restricted to persons with professional qualifications, thus limiting the number of eligible members, Council would need to raise membership dues to counter any shortfall in revenue. There was also a major concern that a smaller group of professionals would not have the same impact on public or governmental forums. The alternative was to foster an industrial society, where anyone associated with the mining industry could become a full member. After all, Mackenzie wrote, "It was a curious fact that each Council, from year to year, places a somewhat different interpretation upon the By-Laws governing election of members."[182]

[181] G.C. Mackenzie, ibid., 26.
[182] G.C. Mackenzie, ibid., 27.

Mackenzie's five-step plan for promoting the Institute as a servant to industry included increasing the awareness of safety and efficiency in mining and metallurgical operations, the conservation of mineral resources, better education for the public, and the stimulation of new investments in mining. Good relations with the federal and provincial departments of mines were vital for improving legislation, along with the intelligent management of each working sector. Finally, companies in the industry needed to practice more cooperation in order to stimulate ore production, treatment, transportation, and exchange. If these proposals were enacted, and the Institute membership reached the "not impossible figure of 5,000," then the CIM could implement a divisional system to represent each provincial territory.

In answer to Mackenzie's article, Mortimer-Lamb, who was personally acquainted with many of the Founders, submitted his own views on the membership question and the purpose of the Institute. In a lengthy and articulate essay in the *Bulletin*, he argued that when the Charter was originally drafted, the wording was intended to mean professional status:

> A majority [of members were] desirous of organizing the Institute on a definitely professional basis, but in the year 1898 the number of professional and technical mining men domiciled in Canada was far too small to make such a course feasible, and the proposal to adopt it was dropped.[183]

He also stated that the Institute had not deviated from its original mandate, and certainly had not developed into "a pseudomorphic form." Furthermore, if the founders of the Institute had intended it to be a servant to industry, they would not have made its first purpose in the Charter "to promote the arts and sciences connected with the economical production of valuable minerals and metals." Regular meetings were still held for technical exchanges, and the Council continued to promote the interests of the mining community through its scrutiny of government policy. If the membership doors were thrown wide open, he warned, the Institute could get dragged into the commercial and financial aspects of industry, areas which it had previously avoided. Consequently, the motives of the professionals, who took an active interest in the *Proceedings*, were different from those of non-professionals, who paid less attention to publications and were driven

[183] H.Mortimer-Lamb, ibid., 399.

ANNUAL INTERPROVINCIAL CONGRESS

— OF —

Canadian Mining Engineers, Mine Managers and Mining Men

. . UNDER THE AUSPICES OF . .

THE CANADIAN MINING INSTITUTE

. . . WILL BE HELD IN . . .

THE CLUB ROOM, WINDSOR HOTEL, MONTREAL

ON 1st, 2nd, and 3rd MARCH, 1899.

Sessions Commencing
10.30 a.m., 3 and 8 p.m.

Lantern Projections
Thursday Evening.

THE FOLLOWING WILL CONTRIBUTE PAPERS FOR DISCUSSION AT THESE MEETINGS

DR. JAMES DOUGLAS, Copper Metallurgist, New York.
DR. GEORGE M. DAWSON, C.M.G., Director Geological Survey of Canada.
MR. A. R. LEDOUX, Metallurgist, New York.
MR. BERNARD MACDONALD, M.E., Salmon River, N.S.
MR. JOHN BIRKINBINE, M.E., Philadelphia, Pa.
DR. HENRY AMI, Geological Survey of Canada, Ottawa.
MR. A. McCALLUM, Peterborough, Ont.
MR. F. CIRKEL, M.E., Vancouver, B.C.
MR. F. T. SNYDER, Peterborough, Ont.
MR. A. H. HOLDICH, Nelson, B.C.
MR. J. T. DONALD, M.E., Montreal.
MR. A. HOLLAND, Ottawa.
MR. A. J. COLQUHOUN, Savonas, B.C.
MR. WM. BLAKEMORE, M.E., Fernie, B.C.
MR. F. HILLE, M.E., Port Arthur.
MR. JOHN E. HARDMAN, S.B., M.E., Montreal.
PROF. W. G. MILLAR, Kingston, Ont.
MR. C. A. MEISSNER, Londonderry, N.S.
MR. CHARLES FERGIE, M.E., Westville, N.S.
PROF. DeKALB, Kingston, Ont.
MR. JOHN B. HOBSON, Quesnelle, B.C.
MR. R. G. McCONNELL, Geological Survey of Canada, Ottawa.
MR. J. BURLEY SMITH, M.E., Winnipeg.
MR. JAS. D. SWORD, M.E., Rossland, B.C.
MR. R. C. CAMPBELL JOHNSTONE, M.E., Nelson, B.C.
MR. JAS. F. LEWIS, Chicago, Ill.
MR. D. SMITH, Kingston.
MR. GEORGE E. DRUMMOND, Montreal.
MR. R. W. BROCK, B.A., Ottawa.
MR. W. HAMILTON MERRITT, A.R.S.M., Toronto.
MR. J. W. EVANS, M.E., Trenton, Ont.

REDUCED RAILWAY RATES

By special arrangements made with the Railway Companies mining men attending these meetings will be carried to and from Montreal at reduced fare as under.

From points on the—INTERCOLONIAL RAILWAY—SINGLE FARE
" " QUEBEC CENTRAL RAILWAY "
" " CANADA ATLANTIC RAILWAY "
" " CANADIAN PACIFIC RAILWAY "
" " GRAND TRUNK RAILWAY "

In order to obtain these reduced fares members and their friends and all who intend being present at the meetings must obtain from their Ticket Agent a Certificate which must be signed by the Secretary of the Institute at the meetings.

Everyone interested in Mining or Metallurgy and in the Development of the Mineral Resources of the Dominion is invited to be present and take part in the proceedings of these Meetings.

ANNUAL DINNER

Members and their friends will dine together in the Windsor Hotel, on Friday evening at eight o'clock.

JOHN E. HARDMAN,
PRESIDENT.

B. T. A. BELL,
SECRETARY

S.S. Fowler, 1900-01
London & B.C. Gold Fields Ltd.,
Nelson. B.C.

Charles Fergie, 1901-03
Dominion Coal Co., Glace Bay,
Nova Scotia

Eugène Coste, 1903-05
Consulting Mining Engineer,
Toronto

George R. Smith, 1905-07
Bell's Asbestos Co.,
Thetford Mines, Quebec

Four early Presidents of the Institute

Officers and Council. Canadian Mining Institute. 1906-07

(1) Frank B.Smith (Council), Edmonton, Alta. (2) J.C. Gwillim (Council), Kington, Ont. (3) Jas. McEvoy (Council), Fernie, B.C. (4) Geo. R. Smith (President), Thetford Mines, Que. (5) Harry Williams (Council), Danville, Que. (6) A.M. Hay (Council), Toronto, Ont. (7) W.G. Miller (Council), Toronto, Ont. (8) R.R. Hopper (Council), Montreal, Que. (9) C.J. Coll (Council), Stellarton, N.S. (10) H. Mortimer-Lamb (Secretary), Montreal, Que. (11) S.S. Fowler (Past President), Nelson, B.C. (12) Chas. Fergie (Past President), Glace Bay, N.S. (13) A.E. Barlow (Council), Ottawa, Ont. (14) J. Bonsall Porter (Council), Montreal, Que. (15) Thos. Cantley (Council), New Glasgow, N.S. (16) E. Coste (Past President), Toronto, Ont. (17) F.D. Adams (Vice President), Montreal, Que. (18) Frederic Keffer (Council), Greenwood, B.C. (19) D.W. Robb (Council), Amherst, N.S. (20) John Blue (Council), Eustis, Que. (21) R.G. Leckie (Vice President), Sudbury, Ont. (22) J. Stevenson Brown (Treasurer), Montreal, Que. (23) G. Herrick Duggan (Vice President), Glace Bay, N.S. (24) B.A.C. Craig (Council), Toronto, Ont. (25) Thos. Kiddie (Council), Victoria, B.C.

Members of the Institute's 1908 Summer Excursion at the famous Lawson silver mine in Cobalt.

The Institute's 1908-10 President, Willet G. Miller, at the Trail smelter of Consolidated Mining and Smelting Company (later Cominco) during the 1908 Summer Excursion.

1908 Summer Excursion: "How coal men inspect metal mines. Mr. Fergie and Mr. Lindsay, at Phoenix [copper mines], discussing the 'Future of Coal Mining in Canada'." Charles Fergie was President of the Mining Society of Nova Scotia in 1898-1900, and Institute President in 1901-03.

D.H. Browne of the Canadian Copper Company (later Inco) conducting members of the 1908 Summer Excursion through the smelter at Copper Cliff.

A.E. Barlow (top) succeeded Frank Dawson Adams (bottom) as CIM President in 1912-14. They had worked together on the geology of the Precambrian as officers of the Geological Survey of Canada. Barlow's career was tragically cut short when the Empress of Ireland sank in the Gulf of St. Lawrence in 1914. Adams went on to become Logan Professor of Geology at McGill University, and an eminent earth scientist and teacher. (Barlow photo, Geological Survey of Canada 202385)

Lieutenant-Colonel John J. Penhale was awarded the Distinguished Service Order for his conspicuous service with Canada's First Division in France during the Great War. One of the fathers of the asbestos industry in Quebec, the genial Colonel was a CIM Founder and a member of the first Council. His interest in the Institute never flagged, and while in France he organized the unofficial but highly successful Flanders Branch of CIM.

CANADIAN MINING INSTITUTE BULLETIN, MAY, 1916 439

Exhibit "E."

B. E. F.

FLANDERS BRANCH

CANADIAN MINING INSTITUTE

FIRST (and we hope the last) Annual Dinner

IN THE DUG OUT

37 Grande Place, SOMEWHERE IN FRANCE

FIELD CENSOR PASSED 801

THURSDAY EVENING

MARCH 2nd., 1916.

Ne vous méfiez pas!

Des oreilles amies vous écoutent.

Carriages at 10.30 (Gott strafe the Town Major)

MENU

WHAT WE ARE TO HAVE

SOUP
Desecrated Vegetable

FISH
Sardines à la B. E. F. Canteen

ENTREE
Maconochie Ration
(Multum in Parvo)

GAME
Jugged Hare
(if D. R. O. 1225 is cancelled)

JOINT
Bully Beef à la C. A. S. C.

SWEETS
Marmalade Biscuit
Chlorinated Water
Ration Rum (perhaps)

WHAT WE WOULD LIKE TO HAVE

POTAGE VELOUTE

VOL AU VENT
Aux Champignons

POISSON
Sauce Verte

FILET DE BOEUF
Légumes
Asperges en Branches
Sauce Mousseline

VOLAILLE TRUFFEE
Aux Petits Pois

FRUITS

DESSERTS

CAFE

by self-interest. Mortimer-Lamb's suggestion was to divide the Institute into two sections, one technical, the other industrial. The former would represent the professional members of the Institute, under the designation of "Fellow," whereas the latter could function along the lines of a "Canadian Mining Association, or Chamber of Mines." Some years later, the industry itself would organize a Canadian Mining Association, now known as the Mining Association of Canada.

The Divisional Proposal

The concept of divisional representation was formally introduced by O.E.S. Whiteside, the 1920-21 Institute President, at the Annual Western Meeting in Winnipeg in 1921. A member since the beginning, he had been on Council for two terms, and was one of the original organizers of the Rocky Mountain Branch. As Secretary-Treasurer of the British Columbia Division, Mortimer-Lamb supported Whiteside's proposal. The great distances between mining centres in British Columbia, plus the diversity of occupations, made it hard to recruit new members, and an eastern Headquarters compounded the difficulties as few members could attend national events.

Mackenzie took the idea one step further, and proposed the formation of four divisions: Atlantic (Nova Scotia, New Brunswick, and Prince Edward Island), Eastern (Quebec and Ontario), Western (Manitoba, Saskatchewan, Alberta, and Northwest Territories), and the Pacific (British Columbia and Yukon). In addition to the dues paid by an active membership, revenues could be augmented by the attraction of corporate subscriptions, and advertising, and by the sale of publications. Each division would share the total revenue in proportion to its membership, and a centralized headquarters would employ a permanent administrative secretary. Members would elect five or more councillors from each division, who would also serve as Institute councillors. These councillors would control and direct local matters, but instead of answering directly to Council, they would fall under divisional jurisdiction.

It was clear that Mackenzie had given this scheme considerable thought because he also presented the opposing views: that the need for a controlling Headquarters might be reduced; that the Institute might split into smaller parts; or that an organization, defined by professional interests, might supersede those of industry. According to him, these fears were completely ungrounded:

> The divisional system could not be expected to bring about the dissolution of the parent body, because several members would quickly realize that their existence as co-partners in the parent national organization is their one guarantee of industrial and financial strength. . . By cutting themselves off from the central body they would, in truth, be selling their birthright for a mess of pottage.[184]

The divisional plan was accepted with modifications, and in January 1922, a list of Officers and Council, printed in the *Bulletin*, outlined not four, but six geographical regions: Nova Scotia, Quebec, Ontario, Manitoba, Alberta, and British Columbia. Each region had five Councillors, and with the President, six Vice Presidents, and three Past Presidents, the total number on Council was forty.[185] These divisions would be redefined as Districts in 1945.

A Question of Qualifications

The controversy over membership qualifications continued to rage throughout the spring and summer at meetings and in the *Bulletin*. In an article published in November 1922, Mackenzie insisted that "the Institute must possess a distinctive policy if its functions and activities are to be apparent and easily understood by those whom it desires to enrol as members."[186] At the annual meeting in Ottawa, the majority favoured the welcoming of anyone with mining interests, and the non-technical would share equal status with the technical. When Mackenzie asked members of the Mining Society of Nova Scotia what they thought, they admitted to having had an on-going discussion and debate on that particular topic for the past thirty years. One prominent member stated that:

> If the Society had been only a professional organization, then people like himself and a Past President, D.H. McDougall, would have been barred. . . we have both graduated from the school of experience, and then took a post-graduate course at the University of Hard Knocks.[187]

---

[184] G.C. Mackenzie, ibid., 31.

[185] "Officers and Council," *Bulletin*, no. 117 (January 1922). In subsequent years, "Manitoba" was replaced by "Manitoba-Saskatchewan", and "Nova Scotia" by "Maritimes".

[186] G.C. Mackenzie, "Institute Policy," *Bulletin*, no. 127 (November 1922): 1135-1155.

[187] Pigot, ibid., 72-73.

On May 25, 1923, the Secretary-Treasurer mailed 1,218 letter ballots to the membership, who were being asked to agree that the Institute would have five classes of members: Honourary Members, Corresponding Members, Members, Associates, and Students. Associates would be able to vote, but not hold office.[188] Despite Council's best efforts to publicize the vote, the returns were disappointing. Only 28.2% of eligible members replied, leading Mackenzie to conclude that the remaining 71.8% were totally indifferent, and the Institute's affairs would continue to be "directed by an active minority."[189] The outcome was uncontested as over two thirds of the ballots were in favour, but even after the amendments became official in August, some people were still unclear about their interpretation. The clause which caused the most concern dealt with new members, who had to be at least twenty-seven, employed for five years in a responsible position in practical mining or metallurgical work, or who, at the discretion of Council, had sufficient technical or scientific knowledge to merit the professional label.

The problem hinged on the word "responsible." Of the complaints printed in the *Bulletin*, H.E.T. Haultain's was the most direct:

> As it stands there are important differences of opinion as to what it really means, as to what it was intended to mean, and as to what it might mean. The only way it can be made workable is for Council to place some interpretation upon it, and I again ask the question: Does the new By-Law on Membership broaden out the membership or make it more restricted?[190]

The debate on who could be a member, and what their professional credentials ought to be, would take much longer to resolve. Even Mackenzie acknowledged that some restraint was necessary because quick changes would not produce the best policies. In fact, this discussion would be postponed for another twenty years. When Canada slowly emerged from the recession in 1924, it is possible that Council's efforts were diverted towards new projects and concerns, especially in how to keep the MSNS from leaving the fold.

---

[188] "Society Affairs: Membership," *Bulletin*, no. 136 (August 1923): 522.
[189] "Notes and Comments," *Bulletin*, no. 134 (June 1923): 477.
[190] H.E.T. Haultain, "The New Membership By-laws," *Bulletin*, no. 142 (February 1924): 40.

A Misunderstanding

Mackenzie's crusade to clarify the policies and By-laws of the Institute did not stop with membership. On September, 15, 1922, he sent a letter to all MSNS members outlining their obligations to the Institute and asking each person to pay three dollars extra to support the *Bulletin* in addition to their regular dues.[191] The letter had completely circumvented the MSNS executive, and had infuriated its President, G.D. McDougall, who argued that it directly violated the special terms of agreement between the Institute and the Society. McDougall also indicated that many MSNS members had been unhappy with the affiliation, and perhaps it was time to re-assess the wishes of the entire body. In the meantime, the MSNS Council drafted its own letter informing its constituency that the matter was under scrutiny, and that no action need yet be taken.

Mackenzie's initiative could have resulted in a complete breakdown in communications had the Institute President, John Dresser, not attended the Society's annual meeting in Halifax the following April. Due to the sensitivity of the issue, most of the negotiations took place outside the boardroom and are not well-recorded. In his official address to the MSNS members, Dresser insisted that "there was no intention on the part of the Institute to do other than to assist them in every way. There could be no progress if any part of the body lagged behind, and the Nova Scotia Society was an integral part of the Institute."[192]

The problem was narrowed down to the Society's chronic shortage of funds, and the lack of motivation to attract new members because for every member over the one-hundred mark who joined the MSNS, the Institute acquired all of the dues. This policy was clearly unfair, and was the ultimate source of aggravation. A committee, which included Dresser and several prominent MSNS members, eventually reached a compromise whereby the Society could retain \$7.50 out of the ten paid by each new member.[193] This seemed to breach the impasse, as no further mention of the issue was made during the year. It had been a close call. The alliance between the Institute and the MSNS was tentative at best, and would always require understanding and diplomacy.

In 1923, the Institute marked its twenty-fifth anniversary, but due to the difficult economic times, the anniversary was observed without a

[191] Pigot, ibid., 73.
[192] Pigot, ibid., 74.
[193] Pigot, ibid. The other committee members were Charles O'Dell (MSNS President Elect), F. Gray, and W.H. Collins of the Geological Survey of Canada.

lavish celebration. As a special tribute, Norman R. Fisher, who would become Institute President in 1929-30, moved at the Council meeting of January 18, 1924, that all living Charter Members be elected Life Members. These men were William A. Carlyle, Charles Fergie, Frank Loring, and John J. Penhale. Another man, William Nichol of Kingston, was also recognized for his able contributions to the science of mineralogy and his devotion to the mining students at Queen's University. Although still living, John Hardman was not included in this list, and there was no reason given for his omission.

The Question of Divisions Again

On December 27, 1929, Mortimer-Lamb posted a note to Mackenzie countering his statement that there were no Council records authorizing the establishment of the British Columbia Division. As sharp as ever, Mortimer-Lamb was certain that it had been approved ten years previously. On his urging, Mackenzie sifted through the archives and found an item dated February 17, 1920, which asked that the name of the British Columbia Branch be changed to the British Columbia Division.[194] However, it appears that even though Councillors had been favourable to the term "Division," they had not considered any new provisions in the By-laws, nor did they take any action to remedy the situation by securing advice, legal or otherwise. This reopened the question of divisions on March 5, 1930, and placed it back on the agenda for the next annual meeting.

A special committee was appointed to investigate the definition of Branches, Sections, and Divisions, their relationship to the Institute, their means of organization, and their possible financial assistance from Headquarters. At the AGM in 1930, a new by-law was proposed by the committee, which gave the Council broad leeway:

> The Council may, from time to time, make such arrangements as it may deem fit for the establishment and maintenance of regional or technical Divisions and local Branches or Sections of the Institute to embrace the membership of any particular region, profession or dis-

[194] CIM, *Council Minutes*, April 25, 1930, p. 50, vol. 4, NAC. Although a divisional structure had been implemented in 1922, the term "Division" appears to have been formally applied only to the British Columbia Division.

> trict. These Divisions, Branches, or Sections shall be governed by such regulations as may be made from time to time by Council.[195]

This became the starting point for any serious consideration of specialized technical groups; although small, technical sections had already been around for several years. In 1926, Mackenzie had announced that several oil men in Calgary, Toronto, and eastern Canada were initiating plans for an Oil Section, and *Council minutes* indicate that the news was received with "warm approval."[196] Consisting of forty to fifty members, the section was officially formed in Calgary on November 29, 1926, and called the Petroleum and Natural Gas Section. Like the other technical sections in existence, it would function like a branch, its members would pay no additional fees, and it would submit relevant papers and discussions to the *Bulletin*. At its initial meeting, the new section adopted several by-laws of the AIME's Petroleum Section, and then presented them to Council for approval.[197] Twenty-odd years later, some of these by-laws would challenge some of the best minds of both organizations.

End of the Recession

In the mid to late 1920s, the Canadian economy rebounded due to investment in old and new staple industries. While the coal and iron ore sectors had both prospered and declined in various parts of the country, the demand for domestic appliances and other consumer goods created new markets for metals. In western Canada, renewed demands for oil and natural gas had led to the discovery in 1920 of oil at Norman Wells on the Mackenzie River. This augmented the reserves in the Turner Valley of Alberta, and held out the promise that more deposits could be found. Grain farmers expanded their fields in response to the foreign demand for wheat, and Canada's newest commodities, pulp and paper, were destined to top the world's market.

Improvements in rail and air transport were particularly important to the surge in development. An increased network of railway lines opened parts of the country that had previously been inaccessible to most people.

---

[195] CIM, *Council Minutes, Report of Committee on Divisions and Branches*, 1930, p. 62, vol. 4, NAC.
[196] CIM, *Council Minutes*, March 22, 1926, p. 332, vol. 4, NAC.
[197] Ibid., 339.

Prospectors took advantage of the inexpensive passenger fares to explore new terrain, and older, less profitable mines were suddenly within reach of a dependable freight system. H.G. Nichols, at the Seventh Annual General Meeting of the British Columbia Division, stated that:

> There has been more practical development of prospects, and there have been more old mines re-opened during the past year or eighteen months, than has been the case at any time since the early days of mining in British Columbia.[198]

Prospecting was also rapidly becoming airborne as planes were used for mineral and oil exploration. An article in the 1919 *Bulletin* on copper deposits on the Coppermine River and the Bathurst Inlet discussed the advantages of air service. Destinations that used to take two months of trudging through the Arctic were now accessible within several hours.[199] In 1921, the Imperial Oil Company began to fly two Junker 185s to Norman Wells, the first in a series of bush planes that would serve the company's northern outposts.[200] Within five years, the use of aircraft was almost routine in spite of the risks, and contributed to the development of Noranda's Horne mine at Rouyn, Quebec, in 1924, and the gold deposits at Red Lake, Ontario, in 1925.[201] With aircraft, remote locations could be supplied throughout the long winter. As the decade closed, William "Bill" G. Jewitt's aerial exploits would lead to the Consolidated Mining and Smelting Company's rich Con gold mine at Yellowknife, and accelerate the development of its lead-zinc mine at Pine Point on the southern shore of Great Slave Lake.[202]

Prior to World War I, technology had been viewed by government and private industry as a short-term economic solution that was not integral to resource development.[203] This attitude gradually changed as companies such as the Consolidated Mining and Smelting Company (later

---

198 H.G. Nichols was Chairman of the Pacific Great Eastern Railway. *Bulletin*, no. 168 (April 1926): 404-413.

199 J.J. O'Neill, "Deposits of Native Copper in Arctic Canada," *Transactions*, vol. XXII (1919): 406-416.

200 Peter McKenzie-Brown, Gordon Jaremko, and David Finch, *The Great Oil Age: The Petroleum Industry in Canada* (Calgary: Detselig Enterprizes Ltd., 1993) p. 36-44.

201 Zaslow, ibid., 359.

202 W.G. Jewitt was a test pilot during World War I, and a CIM member. He was inducted into the Canadian Aviation Hall of Fame in 1978 for his role in northern exploration. W.G. Jewitt, "The Mine Makers of Pine Point," *Cominco Magazine* (April 1966). "Cominco Pioneer Aviator to Hall of Fame," *Nelson Daily News*, January 25, 1978, p. 5.

203 Jeremy Mouat, "The Assistance of Science and Capital: The Role of Technology in Establishing B.C.'s Hard Rock Mining Industry, 1876-1906," *Scientia Canadensis*, vol. XVI, no. 2 (1992): 165-183.

Cominco) began to reap the benefits of their own metallurgical research. With a large component of academically trained men, the emphasis on applied science made tremendous gains. From 1917 to 1920, Randolph W. Diamond, who was to become CIM President in 1948-49, and his five-man team at Trail, B.C., developed a radically new procedure for treating the complex Sullivan mine ore. The result was one of the world's first successful large-scale differential flotation operations. Although this solved a difficult technical problem, the larger dilemma for new mining projects was financing.

Much to the dismay of some Institute members, most of the financial backing needed for new mining projects came from foreign sources. Believing that Canada was about to lose control over its domestic resources, O.E.S. Whiteside (CIM President in 1920-21) urged members to educate the public about the rewards of investing in their own country:

> The biggest and the best mines in Canada are not owned by Canadians. It is almost hopeless to attempt to raise capital in Canada for home-mining adventures, however worthy. Why? Because the Canadian public has been led to believe that mining is a gamble. . . The majority indeed are unable to realize that to invest money in the development of a promising prospect is an entirely different thing from speculating on the margin or otherwise in mining stocks.[204]

Manitoba's Flin Flon mine was a prime example. Discovered in 1916, the deposit could not be exploited because of technical problems. These could be solved given enough time, but money was the real obstacle. At the time, the only Canadian company with enough experience to attempt development was The Mining Corporation of Canada, and it lacked the capital for a multi-million dollar facility. Instead, controlling interest was acquired by the Whitney Group of New York in 1925. It had the means to invest in a flotation mill and a cyanide plant, and in 1928, constructed a concentrator, copper smelter and zinc plant, the latter of which was operational by 1930.[205] Although American-controlled, the Hudson Bay Mining and Smelting Company operations at Flin Flon would provide many jobs for skilled Canadians.

---

[204] O.E.S. Whiteside, "The Institute and its Functions," *Transactions*, vol. XXIV (1921): 14.
[205] LeBourdais, ibid., 235-242.

The Great Depression

Just as people were starting to regain some of the confidence they had lost earlier in the decade, the crash of the New York stock market on October 29, 1929, altered their perspectives and ruined their prosperity. The subsequent decline in trade affected transportation, banking, commerce, and industry, and caused mass unemployment. Conditions in the Atlantic provinces were especially difficult, and on the prairies, the worst drought in living memory resulted in farm foreclosures and bankruptcies. In the mining industry, there was a marked drop in the overall value of mineral production. In the *Bulletin*, the Institute reported a drop of 18.6% in just one year, or $279,873,578 in 1931 compared to $227,769, 899 in 1932.[206] In the same year, the price of copper hit an all-time low of 5¢ per pound. However, the blow to the mining industry was softened when the United States government dramatically increased the price of gold from $20 to $35 per ounce. This stimulated the opening of many new gold mines in Canada. Among them were the Lamaque mine, which became the largest straight gold producer in Quebec, the Little Long Lac and Kerr-Addison mines in Ontario, and the Bralorne mine in British Columbia, where the venerable Nickel Plate mine also resumed operations.

Like mining, the pulp and paper industry helped ease the job situation, but fisheries, lumbering, and steel production suffered greatly. With over 1.5 million Canadians out of work, the Depression witnessed long lines at soup kitchens, frequent suicides, and vagabonds atop railway cars.

Improvements at Headquarters

Before the Depression, the Institute had been experiencing another surge in membership and Mackenzie was starting to wilt under the demands of bookkeeping. In 1925, he asked permission to either buy a Kardex system, or hire someone to assist with clerical duties. The Kardex was estimated to cost $400, and Council agreed to the purchase rather than to add to the staff. Another problem was that the space shared by the office and library was again inadequate. In the library, shelf space for new journals, and information and catalogues on mining machinery and supplies was at a premium. Because of the expense and the extra work involved in hosting the forthcoming Second Empire Mining and Metal-

[206] CIM, *Council Minutes*, 1932, p. 113, vol. 4, NAC.

lurgical Congress in 1927, a decision to relocate was deferred. Then in 1928, Headquarters had the opportunity to move from the 6th floor of the Drummond Building to the 9th, which enabled the offices and library to settle in larger quarters.

Second Empire Mining and Metallurgical Congress

In 1927, the Institute hosted the Second Empire Mining and Metallurgical Congress. The first congress, held three years earlier in London, England, had brought together the institutions representing the scientific and technical interests of the mineral and metal industries of Great Britain. It was during this congress that an Empire Council of Mining and Metallurgical Institutions was created to serve as a forum for communication and technical exchange throughout the Empire. This would be achieved by holding congresses at regular intervals.[207] When the Institute was invited to join the Empire Council it readily accepted.[208] Evidently sensing an opportunity to raise its own profile, and to showcase Canada's mineral industry, the Institute offered to convene the second Congress in 1927. The invitation was accepted. In a letter to the British delegates, Canada's High Commissioner in London, P.C. Larkin, wrote:

> Those responsible for the organization of the Empire Mining and Metallurgical Congress could hardly have chosen a better time for a visit to Canada. . . Already the third among our primary industries, mining in the Dominion is now recognized to be on the verge of an epoch of expansion.[209]

The Congress took the form of technical meetings and industrial tours across Canada, a pattern that was to be repeated in future congresses. The *Bulletin* reported that the Congress was a "wonderful success, and it has amply demonstrated, if that were necessary, the great value of such gatherings, not only as they are concerned with the more efficient development of our mineral resources and the industries dependent on them, but in their wider Imperial significance." And with typical

---

[207] "Empire Congress and Council of Mining and Metallurgical Institutions," *Transactions*, vol. XXVII (1924): 5-7.

[208] "Empire Congress and Council of Mining and Metallurgical Institutions," *Bulletin*, no. 142 (February 1924): 35-36.

[209] "High Commissioner for Canada Sends Greetings", *Bulletin*, no. 185 (September 1927): 973-974.

Canadian modesty, added, "We in Canada therefore feel that, in having had the privilege of acting as host to the Congress, we have received more than we have given."[210]

In the interim, interest and involvement in CIM affairs had been extremely satisfying. In 1929, there were six active Standing Committees: Executive, Finance, Library, Publications, Barlow Memorial Prize, and Student Cash Prize. For the first time in its history, the Annual General Meeting was held west of Toronto, in Winnipeg. It was also the third straight year that Council reported an increase in the membership. In a surprise move, Council finally decided to offer rebates to Branches on an experimental basis in order to stimulate their activities and increase local interest. Unfortunately, the euphoria of success at Headquarters was about to tumble with the Depression, and was further compromised by the sudden death of its Secretary.

Just as the Institute was about to process applications for the promised rebates, it lost its annual grant from the Dominion government. This had a devastating effect on the Institute's finances. In addition, the list of members who could not pay their annual dues for 1930 had sharply increased, and both resignations and delinquent removals began to chip away at the membership roster. Although revenues had exceeded expenditures by $257 in 1930, by 1931 the Institute had a deficit of $2,120 which represented 14% of total revenues. The *Bulletin* was steadily losing money. In 1932, the Publications Committee would announce that the *Bulletin* was $1,329 in the red due to a drop in advertising contracts. This would be its lowest circulation in six years.[211] For Mackenzie's successor, E.J. Carlyle, it would be the worst time in the Institute's history to assume responsibility.

Arrival of E.J. Carlyle

On August 22, 1931, George Mackenzie suffered a heart attack at the summer home of friends, and passed away at age fifty-four. His much lamented departure came at a time when "the Institute was experiencing severe difficulties due to the industrial depression [which] threatened chaotic conditions in our head office."[212] Council praised him as a faithful, if somewhat controversial figure, who had always acted in the Institute's best

[210] "Hail and Farewell," *Bulletin*, no. 186 (October 1927): 1151-1152.
[211] CIM, *Council Minutes*, September 1932, p. 152, vol. 4, NAC.
[212] CIM, *Council Minutes*, August 25, 1931, p. 105, vol. 4, NAC.

interests. Three days later, Council called a special emergency meeting, and Major George C. Riley volunteered to be interim Secretary-Treasurer for the three months it would take to find a replacement.

Of the four possible candidates, the choice was narrowed down to two excellent men: Sidney Clarence Mifflen and E.J. Carlyle.[213] Mifflen was a Newfoundlander who had obtained a degree in civil engineering at McGill University and, after working in the mining industry, had become Secretary of the MSNS. E.J. Carlyle was older and more experienced, and it probably helped that he was the younger brother of William A. Carlyle, one of the Institute's founders. Known as "Curly" by his colleagues, E.J. had a degree in mining engineering from McGill, but instead of practicing in Canada, he had worked in Montana, Arizona, Russia, Spain, and South America before returning to Canada in 1931. He soon applied for the position at the Institute, and was the successful candidate. His term with the Institute would last for eighteen years, the longest tenure of any Secretary-Treasurer or Executive Director to date. Mifflin, a much younger man, would remain as Secretary of the MSNS for the next sixteen years.

Resignations

One of the first issues confronting Carlyle was membership, and in 1932, the rolls were dismal. Although 1,004 men remained in good standing, another 46 had resigned, 21 had been removed, and 568 were in arrears. Carlyle insisted that, as a large number of men were jobless, they deserved the "most sympathetic consideration" and that "extraordinary efforts should be made at Headquarters to find employment for such members."[214] To help alleviate the situation, the Institute's Employment Bureau was stepped-up in the *Bulletin*, and members were urged to network with possible employers who frequented the local Branches.

For some men the situation was more critical, as evident in a request from James B. Mawdsley to E.J. Carlyle in mid-October, 1931:

> I am sorry I have so far found it impossible to meet my dues. Things have been rather difficult this year and the Institute's moderate fee has been a problem. . . a report on

[213] The other two men were R.W. Brigstocke and A.R. Lawrence.
[214] CIM, *Council Minutes*, March 31, 1932, p. 127, vol. 4, NAC.

> summer work is now about due so that I expect to forward this money within the next weeks or so. If this is not against the constitution I would be much obliged if this delay could be granted.[215]

As Secretary, Carlyle was in the difficult position of upholding the rules while dealing with his own conscience. Mawdsley, a loyal member since 1922, was a geology professor at the University of Saskatchewan. Like other men, he had temporarily fallen on hard times. Six months later, after a second letter from Carlyle had urged Mawdsley to find the necessary money, Mawdsley replied:

> I appreciate fully what you say and realize the Institute needs all the help possible, and I am sorry that it has proved impossible for me to play my part recently. There are many reasons why I wish to be a member in good standing of the Institute, but financial conditions that not only affect the immediate family but others has made it impossible for me to afford my fees. I might say that the few other societies of which I am a member I have to deal with in a same manner. I nevertheless hope that things will shortly improve and that I will be able to again support the Institute, which is the only professional society I belong to (or did belong to!).[216]

Fortunately for Mawdsley, their correspondence was enough to keep Carlyle informed of the situation, and enough trust was generated between them to guarantee that the arrears would be corrected given enough time. The Institute eventually benefited by Carlyle's patience, as Mawdsley would become Institute President in 1961-62. Other men, who gave up under the stress of financial difficulty and did not communicate their situation, were quietly deleted from the membership rolls. Many of them were completely lost to the Institute as they could never afford to be reinstated.

Throughout the Depression, men who found themselves in positions to assist younger colleagues often went to extraordinary lengths.

[215] J.B. Mawdsley, letter to E.J. Carlyle, October 15, 1931, vol. 4, NAC.
[216] J.B. Mawdsley, letter to E.J. Carlyle, May 23, 1932, vol. 4, NAC.

John A. Allan, Institute President in 1932-33, never lost sight of his students. As Professor, Chair, and founder of the Department of Geology and Mineralogy at the University of Alberta, he and his colleagues tried every possible means to place students in jobs during the "dirty thirties". Retired Professor C.R. (Charlie) Stelck recalled that Allan had personally driven him over to the offices of an oil company, walked him into a room, and presented him to the directors as "Here's your man."[217]

Progress at Headquarters

Carlyle's first year as Secretary-Treasurer must have been hectic and stressful. At a Council meeting in December, he reported that the "office staff was seriously handicapped by inadequate and obsolete equipment, especially the addressograph and typewriter."[218] In addition, the collection of books, periodicals, pamphlets, reports, and maps in the Institute library were not indexed, and some members confessed that to save time, they actually wrote to other libraries to ask if the Institute library had the desired article. Because the Institute library duplicated much of the literature found at the Geological Survey library, the most economical way of proceeding was to hire a typist to copy its indices and cross-references. Later, a trained librarian was hired to index the *Transactions*, and an Industrial Arts Index, and an Engineering Index were also purchased.

As the effects of the Depression gradually eased, the membership began to rebound. In 1935, Council received a letter from the Honourable J.E. Perrault, Quebec's Minister of Mines, awarding the Institute an annual grant of $1,000. Then on April 1, 1937, Headquarters was authorized to increase staff, and raise the salaries of the two office "girls", after the rolls had increased to a record high of 2,600 men.[219] In the meantime, three more Standing Committees were added, bringing the total to nine: a *Bulletin* Committee in 1931, a committee to adjudicate the Randolph Bruce Medal, and in 1932, a Committee on Publicity and Public Relations. After these efforts, it must have been very gratifying for CIM members to hear that the Executive Committee had accepted an invitation by the Royal Society of Canada to become an affiliated society.

[217] "Re-dedication," *Bulletin*, no. 877 (May 1985): 16.
[218] CIM, *Council Minutes*, December 31, 1931, p. 122, vol. 4, NAC.
[219] Misses M.A. Young and E.M. Cartledge each received an increase of $125 per month. For statistics on membership, see *Bulletin*, no. 679 (November 1968): 1273.

Industrial Minerals Section

The idea for organizing more formal technical sections under Institute policy guidelines began in 1932, and set the standard for what were to become the technical Divisions. The Industrial Minerals Section would be the first to offer its own technical session at the Annual General Meeting, and a special committee was drafted in 1932 to stimulate interest and arrange for suitable papers.[220] The scope of the new section embraced non-metallic minerals, clay products, and structural materials. What made the Industrial Minerals Section attractive and unique, and set it apart from the older coal, metallurgy, and petroleum and natural gas sections, was the guaranteed publication of articles in the monthly *Bulletin*, a privilege for which the other technical sections would soon press the Institute for equal terms.

The Membership Issue Returns

In 1938, the debate over membership qualifications re-emerged. Seemingly little had been solved by the By-law amendments of 1922, and correspondence between members and Council on this contentious matter was increasing. Issues ranged from the lack of distinction between members and associates, to charges that the Institute had lost the right to call itself a technical institute. A series of changes to the By-laws had already been both applauded and condemned by members of Council. On the verge of a stalemate, a special committee was appointed to deal with the situation. It consisted of the six most recent past presidents: Cole, Ross, Blaylock, Neilly, Allan, and Gray, but they would not be able to reach consensus. Blaylock thought that "our present By-laws are all right and it is only the interpretation which we have put on them that [has] caused the trouble."[221] Others, such as F.W. Gray, saw inherent problems: "It had not occurred to me until recently that anyone could be a "Member" of the Institute. . . without at the same time being an "engineer" in the same sense that this word is construed by the Engineering Institute of Canada."[222]

From April to November, the group made little progress. In bringing its discussion to a wider forum, it became more aware of the general feeling "That being a member of the Institute means nothing whatever."[223]

[220] CIM, *Council Minutes*, October 15, 1932, p. 166, vol. 4, NAC.
[221] S.G. Blaylock, letter, July 27, 1937, in: CIM, *Council Minutes*, p. 176, vol. 5, NAC.
[222] F.W. Gray, letter, November 1936, in: CIM, *Council Minutes*, p. 174, vol. 5, NAC.
[223] Ibid., 177.

Some committee members wanted to introduce the same rules used by professional engineering societies, but others feared that placing too many restrictions on membership would hinder the Institute and negatively affect the industry. There were also some concerns that if Associates were granted equal status to Members, control of the Institute would leave the hands of technical men.

As one proposal after another was submitted and then discarded, George E. Cole, the Chairman, finally recommended on November 27, 1937, that Council be relieved of its responsibility, and that Branches be given the task of devising standards for membership.[224] This was roughly the same period that Branch executives had asked Council to agree that the "policy of admitting non-members of the Institute into membership in local branches be left in the hands of the local branches to decide individually."[225]

The wrangling lasted until 1942 when a list of possible By-law amendments was printed in the *Bulletin* alongside a fifteen-page discussion of the rationale.[226] The President, Wilbert G. McBride, and Past President Balmer Neilly, had personally undertaken to solve this question, and what emerged after three years of study were six classes of membership: Member, Junior Member, Corporate Member, Associate, Junior Associate, and Corporate Associate. It was proposed that Members and Associates be placed on equal terms, including the right to vote and hold office, and that the term 'Junior' in both cases referred to experience, intent, and responsibility rather than age. The rules over technical credentials were tightened, and members of all classes could now use a special distinction after their names, such as M.C.I.M. and A.C.I.M. At this time, the official abbreviation for the Institute was also changed from C.I.M.M. to C.I.M. When put to a vote, the membership agreed that the improvements in the By-laws were satisfactory.

Of the long and entertaining ramifications of implementing the new By-laws, the only one which caused some agitation concerned timing. With a great many Canadians involved in the war effort at home and overseas, H.E.T. Haultain thought that any amendments should wait another year or two, out of respect for those unable to participate. After all, Council had already withdrawn a previous set of proposals in 1939 because it believed that "In such circumstances, no controversial proposals should be

[224] CIM, *Council Minutes*, November 27, 1937, p. 177, vol. 5, NAC. The *Bulletin* recorded these events somewhat differently.

[225] CIM, *Council Minutes*, Conference of Branch Executives, 1937, p. 78, vol. 5, NAC.

[226] "Proposals to Amend the By-Laws of the Institute," *Bulletin*, no. 362 (June 1942): 221-243.

introduced."[227] Although Haultain's displeasure was duly noted, the revised By-laws were set in motion on August 1st.

World War Two

In 1939, King George VI and Queen Elizabeth became the first ruling British Monarchs to visit Canada. In every town and city, they were greeted by thousands of people who had waited patiently for hours along the parade route. Many Canadians would remember the Royal Tour with startling clarity as it had brought a sense of wonder to many lives touched by hardship and sacrifice.[228] One of the highlights for Institute members happened when the King and Queen went underground at the International Nickel Company of Canada's Frood mine at Sudbury. A photograph of the Royal couple in miner's helmets and long slickers was later proudly displayed in the Institute's council chamber facing the memorial to those who had fought in the Great War.

In Ottawa, a large crowd had assembled to witness the Royals unveiling the new National War Memorial. As they passed the Veterans' Guard, the King and Queen suddenly "disappeared into a sea of blue and brown berets."[229] This simple gesture did much to endear them to the Canadian public, and created positive headlines across the nation. S.G. Blaylock would later comment that:

> the event brought home to us not only what our King and Queen really mean to the Empire, but to the whole world. Canadians making any sacrifice will have the feeling of personal love and respect for our sovereigns added to their sense of duty.[230]

Four months later, on September 10, 1939, Parliament would declare war against Germany, and Canadians would rally to defend the Empire against the enemies of humanity. Despite horrific memories of World War I, most people believed that this was "the right, necessary, and *only* thing to do."[231]

---

[227] "Concerning the Classification of Members," *Bulletin*, no. 333 (January 1940): 10-12.

[228] Robert Collins, *You Had to be There. An Intimate Portrait of the Generation that Survived the Depression, Won the War, and Re-Invented Canada* (Toronto: McClellend and Stewart, 1997) p. 50-51.

[229] Shirley, ibid., 267.

[230] S.G. Blaylock, "Canada's Mining Indutry and the War," *Transactions*, vol. XLIII (1940): 1-6.

[231] Collins, ibid., 54.

Canada was ill prepared for the coming conflict. The four-division Canadian Corps, which had distinguished itself so gallantly in the Great War, had been disbanded. The federal governments of the 1920s and 1930s had avoided the suggestion that Canadian forces should be preserved, and had only retained a few professional officers to train the weekend militia, and air and navy auxiliaries. However, the concept that an officers corps could stay *au courant* with new military technologies and tactics was as faulty as the notion that if Canada ever got involved in another war, its troops would simply follow orders from the British.

In reality, few professional officers had survived the budget cuts of 1935. The Canadian militia, managed by the Permanent Force, had no field uniforms, field equipment, or modern weapons. The Royal Canadian Navy's two best destroyers lacked modern weaponry, and the Royal Canadian Air Force's obsolete aircraft were "little more than a flying joke."[232] The top speed of its Siskin biplanes was only 190 km per hour, as compared to the German Messerschmitts, which flew at more than 550.[233] Canada was not prepared for what the Institute's 1940-41 President, George C. Bateman, would soon call "a mechanized war":

> It is a war to be fought with metals, but also for metals, because the British Empire and the US, between them, control 3/4 of the mineral wealth of the world. The country which can maintain this control, and which can hold out until the full effectiveness of these resources can be brought into play, is bound to win.[234]

The shabby condition of Canada's military did not deter volunteers from enlisting at recruitment centres across the country. During the autumn of 1939, more men offered their services than the army could accommodate. Although their reasons varied from adventure to patriotic duty and the chance for steady employment, no-one rushed into the war with the wide-eyed naivety that typified the start of World War I. Soldiers bound for the second conflict believed in the government's promise that their wartime contributions would be funded by Canadian dollars, and that Canadian units would remain intact under their own officers. The latter decision would

[232] David J. Bercuson, *Maple Leaf Against the Axis. Canada's Second World War* (Toronto: Stoddart Publishing, 1995) p. 13.
[233] Ibid., 13.
[234] G.C. Bateman, address, *Bulletin*, no. 340 (August 1940): 322.

prove very costly in lives lost, as almost one-quarter of battalion commanders were found to be totally inept as instructors or combat leaders.[235]

The war abruptly ended the Depression as all sectors struggled to support the war effort. Government research institutions such as the National Research Council (NRC) opened new areas of investigation such as radar, synthetics, nutrition, improved explosives, and aircraft technologies.[236] This increased activity resulted in short-term openings for a few women scientists, but due to the circumstances, their wider participation in professional societies such as the CIM would have to wait.[237] Factories, such as the Ottawa Car Company which increased its workforce from 50 to 1,800 people in one year, joined other companies to build fighter planes, tanks, guns and warships, and to manufacture a multitude of munitions.[238] Production on a scale not thought possible in the thirties provided employment for the skilled and unskilled, and the resulting labour shortage resulted in the hiring of over 255,000 women into non-traditional jobs.

With few exceptions, women's employment during the Second War did not lead to management positions, or a greater number of opportunities in industry. The government considered them as a temporary labour force, and most returned to their domestic roles or resumed previously held activities. Their efforts did not go unnoticed by the Institute, and at the annual meeting of the MSNS in 1943, their well-being was briefly raised in a paper on the training of supervisory personnel. It was stressed that in plants which hired women, the foremen should not expect women to operate the same equipment designed for men because the result would be fatigue and dissatisfaction. Supervisors were encouraged to acquire a copy of a book entitled *The Psychology of Supervising the Working Woman*.[239]

Institute Members at War

As during the Great War, the Institute assured members that all activities and services would be maintained within reason, and advised that:

> Any person on the roll of membership of the C.I.M.M., shall, while serving with the Active Military or Naval forces

[235] Bercuson, ibid., 19.
[236] Jarrell and Gingras, ibid., 6.
[237] Marianne G. Ainley and Catherine Millar, "A Select Few: Women and the National Research Council of Canada, 1916-1991," Jarrell and Gingras, ibid., 114.
[238] Shirley, ibid., 268.
[239] E.K. Ford, "Training Of Supervisory Personnel," *Bulletin*, no. 377 (September 1943): 516-518.

> of the empire and its allies, be deemed an Honourary member of the Institute and shall, from the date of his enlistment and throughout the term of his military or naval service, be exempt from further Institute subscriptions.[240]

As proof of enlistment, members were asked to remit a photo of themselves in uniform. An internal registration system was soon organized, and the CIM, along with the Engineering Institute of Canada, and the Chemical Institute of Canada, compiled a joint list of scientific and technical personnel for Henry M. Tory, the Director of the Technical Section of the Voluntary Service Registration Bureau at Ottawa. In the first war, 157 CIM members had joined the ranks. Council did not have any idea how many men would volunteer to fight this time, but it believed their numbers would be modest. When the final tally was printed in the *Bulletin* in January 1945, the total was about 16% of the membership, or 439 men.[241]

Those who answered the call-to-arms were largely young, single men, and most were listed as Junior Members. They joined units throughout the military, and served in theatres from northeast Africa to northern Europe. Some, like Charles Bruce Ross and William Gilchrist, served with the Royal Canadian Engineers in Europe; others joined the No. 2 Tunnelling Company bound for Gibraltar. Walter Riva and E.H. (Ted) Caldwell entered the Royal Canadian Air Force, and H.T. Airey and Patrick D. O'Connell chose the Royal Canadian Artillery. With the exception of Airey, who would become a prominent member of the Metallurgical Society, these men would all be elected CIM Presidents.

Of the few members who took part in both wars, C.H.E. Stewart and Major Lindsay Foss deserve special mention. Stewart had been a pilot with the Royal Flying Corps in World War I. After demobilization, he graduated from the University of Toronto as a mining engineer and became a well-known figure in mining camps across North America. Then in 1940, he re-enlisted with the Royal Canadian Air Force as a Wing Commander, and flew more than 300 hours the first year. Promoted to the rank of Group Captain, he lived to recount his many adventures. Foss was not as fortunate. From 1914 to 1918 he had served with the Royal Canadian Artillery and Royal Air Force, then worked as an exploration engineer for Dome Mines Limited, the N.A. Timmins Corporation, and the Lamaque mine. He was

[240] "Members with the Colours," *Bulletin*, no. 344 (December 1940): 453.

[241] The total membership was 2,728. The 439 servicemen include 20 killed in action, and 4 reported missing. *Bulletin*, no. 400 (August 1945): 346-352.

married to Katheryn Carlyle, the daughter of William A. Carlyle, a Charter member of the Institute. When World War II broke out, Foss joined a British Tunnelling Unit and barely escaped the Dunkirk evacuation. Later, while stationed at Gibraltar, he suffered a massive stroke which claimed his life. The news came as a great shock to family, friends, and Institute colleagues.

The Home Front

Many CIM members who were past the age for enlistment, or whose expertise was needed in Canada, held important wartime posts in government bureaus or essential services. They included several future Institute Presidents. George C. Bateman (1940-41) became Metals Controller for Canada, Robert A. Bryce (1942-44), Chairman of the War Metals Advisory Commission, and Alfred L. Penhale (1955-56), Deputy Controller for the Asbestos Industry. John F. Walker (1956-57) was another member of CIM's War Metals Advisory Commission which was formed to advise the Metals Controller; he organized the B.C. War Metals Research Board (later the B.C. Research Council). Another man, Charles C. Huston (1954-55), was Supervising Engineer for the Allied War Supplies Corporation in Montreal. Still others worked on construction projects, like William H. Durrell (1960-61), who supervised the building of Labrador's Goose Bay Airport. Horace J. Fraser (1957-58) represented Canada in Washington, D.C., as Assistant Chief of the Ferro Alloys Branch of the Foreign Economic Administration of the U.S. Government.

Like many leaders of the mining industry, Selwyn G. Blaylock, then head of the Consolidated Mining and Smelting Company (Cominco) at Trail, B.C., had to deal with difficult and unprecedented issues caused by the war. As the production of all lead, zinc, and various chemicals was conscripted, there were serious concerns about "enemy aliens operating a key industry in an isolated community."[242] When Italy entered the war in 1940, the labour force at Trail was largely Italian and many of the men enjoyed hunting or sport-shooting in their leisure time. Because this activity could have been viewed by the government as a national threat, Blaylock met with the workers in their community hall and asked them to surrender their firearms. He used his office as a receiving depot, which took on the appearance of a small arsenal, and personally wrote all the

[242] "Delineations of Destiny. Selwyn Gwillym Blaylock (1879-1945)," *Engineering Digest*, (February 1968): 26-27.

receipts. This measure of confidence shielded the workers from government suspicion, and Cominco was able to keep its operations running full-time. The men were very appreciative, and after the war, many applied for Canadian citizenship.[243]

Wartime Research

The Geological Survey of Canada, the National Research Council, and the Bureau of Mines all played vital roles in wartime research. Since the late 1930s, the GSC had been providing much needed assistance to the mining industry. Its report for 1943-44 stated: "Work was devoted entirely to the examination of and search for potential sources of oil, gas, and strategic minerals and to other projects closely associated with the war effort."[244] At the NRC, a laboratory was established in Montreal to conduct developmental work on a nuclear reactor as part of the Allied research efforts on uranium fission. A pilot plant was later constructed at Chalk River, Ontario, for the production of atomic bomb materials. This facility grew into the Atomic Energy of Canada Limited (AECL) Laboratories. At the Bureau of Mines (previously called the Mines Branch) of the Department of Mines and Resources, scientists conducted classfied research on metals and alloys, radioactive ores, and the recovery of uranium.[245]

One of the results of the National Research Council's wartime research was the development of a new process to extract magnesium from dolomite. Called the Pidgeon Process, it was invented by Lloyd M. Pidgeon in 1941. Pidgeon had been invited to join the NRC in 1937 by Lieut. General A.G.L. MacNaughton, then head of the NRC. MacNaughton foresaw that Canada would need a steady supply of magnesium, which was an essential component of incendiary bombs, and when combined with aluminium, resulted in a light, strong alloy used in the production of aircraft. Pidgeon, a brilliant, young chemist from Markham, Ontario, later commented:

> We had an associate committee on metallic magnesium which was set up with representatives from the Research Council, namely me, and a couple of fellows from the

[243] Richard H. Fish, "100 Years of Smelting at Trail, British Columbia: 1896-1996," *Bulletin*, no. 1006 (January 1997): 92-95. "Delineations of Destiny," ibid.
[244] Department of Mines and Resources, *Annual Report*, 1943-44, p. 15, quoted in Zaslow, ibid., 386.
[245] Ignatieff, ibid., 99.

> Mines Branch (normally responsible for metals) and from the Armed Forces."[246]

Two Canadian gold miners, R.J. Jowsey and W.M. Segsworth, and Thayer Lindsley of Falconbridge, helped to finance the construction of the first magnesium pilot plant at the NRC. When the tests proved successful, a commercial plant was rapidly built by the Wartime Metals Corporation. It was located near Renfrew, Ontario, and was managed by Jowsey's new firm, the Dominion Magnesium Company Limited. The Pidgeon Process was a remarkable Canadian achievement, and the manufacturing rights were almost immediately acquired by five U.S. companies for large-scale magnesium production. Newspapers at the time reported such headlines as "Canadian Scientist Speeds the Bombs," "Scientist Helps Win War," and "More Incendiaries For Export." Pidgeon was elected a Fellow of CIM in 1989, and inducted into the Canadian Mining Hall of Fame in 1996.

The Bulletin and the War

From 1939 to 1945, the *Bulletin* was noticeably slimmer, with an average of twenty to forty pages, and its tone reflected the grim times. With a substantial number of members either overseas or fully engaged in war-work, there was little time or incentive to submit papers. The popular "Members at the Front" column was replaced with "In Service to King and Country," but it revealed almost nothing about the hardships on the front lines. Names were listed in connection with official dispatches, but missing were the personal sentiments that had characterized reports in the *Bulletin* during the First War. This was largely due to censorship rules. Although some members received the *Bulletin* while stationed in Britain, most asked that mail be sent in care of parents or spouses. In one of the few personal letters printed in the *Bulletin*, the writer's name was not revealed. On receiving the news that he had been given an honourary membership, and while on maneuvers with the Royal Navy, he wrote:

> I want to thank you for this honour and also the Institute for its thoughtfulness and kindness. Many of our friends are gone now, and personally I feel like a stranger, even in

[246] Robert E. Brown and Ruth Bryson, "Magnesium Industry Legend," *Light Metal Age* (April 1995): 44-46. "Young Canadian Chemist: First Hero in Metals War," *Canadian Business Magazine* (July 1942): 23.

> my own home-town of. . . That is why remembrance by the Institute means even more these days. That letter bucked me up a lot. My parents are keeping the *Bulletins* for me faithfully and I hope that some day soon I will be back using the latest mining methods they suggest.[247]

Even though Institute support was not as visible as during World War I, its pledge, and that of its members, to assist whenever possible never faltered. Gilbert LaBine, President of Eldorado Gold Mines Limited, and a CIM member, donated three grams of radium, or enough for 1500 planes, to the Minister of Aircraft Production for use in illuminating instrument dials during night missions.[248] The British Columbia Division gave an unconditional cheque for $500 to the Honourable J.L. Ilsley, Ottawa's Minister of Finance, an action that was matched by a similar donation by the Winnipeg Branch in 1941.

On rare occasions, the *Bulletin* carried thank-you notes for gifts received. In a letter to Mrs. J.A. Lindsay, President of the Women's Association of the Montreal Branch, the Second Tunnelling Company of the Royal Canadian Engineers gratefully acknowledged the monthly shipments of cigarettes, and 250 Christmas boxes containing socks, blades, soap, and other welcome items. The unit's Commanding Officer wrote:

> No doubt you have seen in the Canadian Press of the safe arrival of this Company back in the U.K. the latter part of last year and we are all very glad to rejoin the Canadian army after a period of almost two years in Gibraltar. . . all the ranks join me with their thanks and appreciation.[249]

Probably the record for assisting servicemen was held by the Mining Women's Association of Toronto. Mrs. J.R. Rea, the President, reported that 12,000 cigarettes, 664 knitted articles, plus hundreds of dollars had been given to the Royal Canadian Engineers, and similar amounts had been shipped to other field units and the Navy.[250]

---

[247] "A Friendly Letter," *Bulletin*, no. 370 (February 1943): 69.
[248] Lord Beaverbrooke was the Minister of Aircraft Production. *Bulletin*, no. 348 (April 1941): 123.
[249] "The Second Tunnellers Thank the Montreal Ladies," *Bulletin*, no. 371 (March 1943): 96.
[250] They had also raised sizeable donations for the Red Cross, the Greek war relief, the British minesweepers, the Toronto Infant's Home, and the Canadian National Railway's travelling schools in northern Ontario. *Bulletin*, no. 371 (March 1943): 96.

CIM carried on with its normal activities, but as expected, the Eighth International Geological Congress had to be postponed indefinitely, and there was troublesome news coming from Great Britain. The offices and library of the Institution of Mining and Metallurgy (IMM) in London had been destroyed during an air raid, but in true British fashion, IMM had still managed to produce its August *Bulletin* on time. Closer to home, a decision had been made by Council to defer granting life memberships in CIM until peacetime, and to retain the committee of five past presidents, which had formerly dealt with membership, to study Institute policy. A short time later, in 1942, Council took steps to add to CIM's technical sections by approving, in principle, the formation of a geology division.

Wartime Committees

Two areas in which the Institute's contributions were exceptionally valuable were on its own Sub-Committee on Conservation, Development, and Reconstruction, and on the federal War Metals Advisory Committee. The former, led by Robert C. Wallace, Principal and Vice Chancellor of Queen's University, and Institute President in 1924-25, identified many of the mining industry's unique problems, and prepared several briefs for the federal and provincial governments. These included ever-present concerns over mining laws, and the importance of the mining industry to the Canadian economy.[251] The Committee was particularly concerned about the viability of mining towns after their mines had been depleted. Mine owners and operators were urged to invest in municipal services and secondary industries so that during lean years, workers could rely on alternative sources of income.

The War Metals Advisory Committee and the need for a Metals Controller began as an Institute initiative. Early in 1940, Council realized that immediate action was needed to ensure an adequate supply of essential resources for the war effort. Members suggested that a Board or Committee be established, headed by "a mining executive of wide experience," and consisting of men who represented all branches of the metals and industrial minerals sectors.[252] George C. Bateman formally placed this recommendation before C.D. Howe, then Minister of Munitions and Supplies:

[251] "Sub-Committee on Conservation and Development," *Bulletin*, no. 368 (December 1942): 587. CIM, *Council Minutes*, December 21, 1942, p.203, vol. 5, NAC.
[252] "The Institute Pledged to Co-Operation," *Bulletin*, no. 338 (June 1940): 259.

> CIM will mobilize the entire resources of the Canadian mining industry behind the war effort to ensure adequate supplies of metals and industrial minerals. [It will] collect and coordinate information regarding possible sources of supply & productive capacity [and] provide a central agency through which all war enquiries relating to metals and minerals [should] be handled.[253]

Howe recognized the proposal's value, and promptly appointed Bateman as Metals Controller on June 18, 1940. Under the authority Canada's Wartime Prices and Trade Board, Bateman was placed in charge of all minerals, ores, metallic products, metals and alloys, except for coal, solid fuels, and iron and steel. He not only had the power to deal in metals and minerals, but to determine market values, limit quantities, or prohibit their sale or distribution. His experience and energy made him an excellent choice for this position. A graduate in mining engineering from Queen's University, he had worked in several different facets of the mining industry across North America, and had served for eighteen years as Secretary-Treasurer of the Ontario Mining Association.

As part of the many-tiered format of the Metals Control Bureau, the War Metals Advisory Committee, which advised Bateman on strategic metals, was composed almost entirely of prominent CIM members.[254] Their efforts, along with the larger Wartime Prices and Trade Board, enabled Canada to avoid the terrible recession which had crippled the economy after the first war. In nearly every commercial or private business, government controllers, regulators, and advisors kept a tight rein on production, rationing, and purchasing. Corporate and personal incomes were taxed, and the sale of Victory Bonds, along with careful regulation by the Bank of Canada, allowed the country to pay for its own military expenses. These and other visionary measures, which aided post-war recovery, were put into action long before there was any real indication of an Allied victory.

[253] G.C. Bateman, letter to C.D. Howe, May 20, 1940, "The Institute Pledged to Co-Operation," *Bulletin*, no. 338 (June 1940): 259. Howe, who was born in the U.S., had a very successful career in Canada. He held eight federal government portfolios from 1935 to 1957, and was a strong proponent of Canadian industry and the privatization of wartime companies.

[254] "Metals Control in Canada," *Bulletin*, no. 368 (December 1942): 573-583. "The Institute and the War Effort," Report of Council for 1942, *Transactions*, vol. XLVI (1943): vi-vii.

CIM Men in Uniform.

Although the Canadian armed forces had meagre beginnings and suffered heavy losses due to inexperience, antiquated equipment, and poor leadership, by 1944 they had emerged as a proud and capable army. As in any major conflict, the cost in human lives was high. Of the 1.1 million men and women who served, 42,042 died, and 54,414 were wounded.[255] On a smaller scale, the Institute's honour roll in 1940 listed thirty-four men on active service. The next year, this figure jumped to 136 with two missing and another killed in action in Egypt.[256] By 1942 the figure reached 212. At the height of the conflict, there were 439 CIM men in uniform, and eighteen would make the ultimate sacrifice.

The war affected almost everyone. For every soldier on active duty, there were parents, spouses, children, and friends who prayed for their safe return. While waiting several anxious weeks for news of his eldest son, Randolph Diamond received a telegram that the plane of his son, Lieutenant J. E. Diamond, had been spotted in the North Sea with no survivors. Another CIM junior member, Flying Officer Lawrence S. Hill of Lucky Lake, Saskatchewan, crashed in dense fog only ten miles from the British coast after a successful bombing mission. Of the many battles which claimed Canadian lives, there is little doubt that Dieppe was the worst. Of the 4,963 officers and enlisted men who landed on the stoney beach on August 19, 1942, only 2,210 returned. Of those that remained in France, 807 were killed in action, 100 died later of their wounds, and 1,874 were taken prisoner.[257] One such man was Lieutenant J.E.R. Wood, a CIM junior member serving with the Royal Canadian Engineers. He was first reported missing, but later found in a prisoner-of-war camp.

End of the War

The German army that had occupied the northern front surrendered on May 5, 1945. Three days later, the formal truce was signed in Rheims, on VE Day, ending the war in Europe. Canadians were still suffering heavy casualties among the ground forces until the very last moment. Their next assignment was to prepare for the Allied assault on Japan, but fortunately for Canada, their assistance would not be required.

---

[255] Bercuson, ibid., 273.
[256] John W. Dallamore, Squadron Leader, Royal Air Force, was a CIM junior member.
[257] Bercuson, ibid., 61, 73.

The steps which led to the construction and deployment of humanity's fiercest weapon have been well documented. But at the time, only a few people knew that a portion of the uranium used for the first nuclear bomb had been mined near Port Radium in the Northwest Territories, or that concentrates were flown fifteen-hundred miles to a uranium refinery at Port Hope, on Lake Ontario.[258] Of the two refineries in the world capable of producing uranium, the other was located in occupied Belgium, making the existence of the Canadian facility a well-guarded secret. Another little-known fact was that a special plant had been built by Cominco at Trail to supply heavy water for the nuclear program.[259] Few would know until it happened that a nuclear device could instantly destroy a large city and kill tens of thousands of people. The annihilation of Hiroshima and Nagasaki ended the Second World War when Japan surrendered on August 14, 1945. In less than three weeks, the Cold War began between the Soviet Union and the western Allies. Scientists, engineers, and politicians had exchanged their quest for peace for the uncertainties of the atom.

In an address to the Montreal Branch of CIM on the development of nuclear technology, Lew Kowarski, a physicist who had worked alongside Frederic Joliot-Curie, Irène Joliot-Curie, and Hans Halban at the Curie Laboratory in France, commented:

> We have every reason to believe that atomic power will someday be used for purposes other than blind and uncontrolled destruction. . . We cannot see at present how it can be done. Uranium as a source of useful power presents some tremendous disadvantages.[260]

When the war ended, the Institute did not create a formal memorial to the members who had served. This was partly due to the deficit caused by absorbing hundreds of complimentary subscriptions to the *Bulletin*, and partly because most members wished to resume their lives as quickly and painlessly as possible.[261] While men like Charles Camsell, CIM President in 1947-48, helped to reorganize postwar government science,

[258] Smith, ibid., 286-287. See also Fathi Habashi, "Fifty Years of Atomic Bombs," *Bulletin*, no. 995 (November/December 1995): 97-105. And Robert Jungk, *Brighter than a Thousand Suns: A Personal History of the Atomic Scientists* (San Diego: Harcourt Brace, 1970).

[259] R.H. Fish, "100 Years of Smelting at Trail, British Columbia: 1896-1996," *Bulletin*, no. 1006 (January 1997): 92-95.

[260] L. Kowarski, "Atomic Power," *Bulletin*, no. 404 (December 1945): 682-689. His affiliation was listed as the NRC.

[261] CIM, *Council Minutes*, June 29, 1948, p. 87, vol. 6, NAC.

the Institute focussed on how to broaden its services, and make membership more attractive to the next generation.

Mines Ministers Conference

From 1944 to 1950, the Institute began a series of initiatives starting with the organization of an Interprovincial Mining Committee in 1944, spearheaded by Allen A. Mackay, CIM President in 1944-45. This Committee brought together the provincial mine ministers and key individuals from industry, and permitted a thorough debate of ideas and opinions from across the country. What seemed to surprise the participants was how different the issues in one region could be from those in another. By all accounts, they returned home with a more enlightened view. So successful was this event that it eventually evolved into the on-going Mines Ministers Conference.

Among the Committee's most active members was Alphonse O. Dufresne, Quebec's Deputy Minister of Mines from 1941 to 1960, and CIM's first French Canadian President in 1950-51.[262] Dufresne fostered the growth of Quebec's mineral industry by improving mining legislation and mine safety, and encouraging the development and operation of mines. During his tenure as Deputy Minister, he improved the province's geological maps, and added a number of valuable services for the mining industry such as publishing technical information, carrying out mine and quarry inspections, providing analytical laboratory services, and conducting industrial research. Dufresne was deeply committed to education, and helped establish a Canadian university scholarship program for young Quebecers wishing to pursue careers in mining and metallurgical engineering. This program led to the founding of the École des Mines at Université Laval in 1934, and the promotion of mining engineering at École Polytechnique. In his honour, CIM created the A.O. Dufresne Award in 1977. His contributions were also recognized by the Canadian Mining Hall of Fame in 1992.

New Technical Divisions

Probably the most far-reaching Institute initiative, under the capable guidance of Presidents A.E. Cameron (1945-46) and R.J. Ennis (1946-47), was the formation of five new technical Divisions to provide for the

[262] "Personality Profile: an Interview with CIM Past President A.O. Dufresne," *Bulletin*, no. 839 (March 1982): 52-60.

growing number of members with special technical interests. These were the Coal Division in 1944, followed by the Metallurgy Division and the Geology Division in 1945, the Metal Mining Division in 1946, and the Petroleum and Natural Gas Division in 1949.[263] To avoid confusion, the term Division was now reserved for technical units. The old British Columbia Division was renamed the B.C. Section on October 5, 1945, and the geographic divisions adopted the term District, but it would take time for the new terminology to become familiar.

Although several technical sections had been formed many years earlier, there is little record of their activities. The timing in creating the new technical Divisions was propitious. Canada was on the verge of a post-war scientific boom, one that would touch every sector of society. As many veterans took advantage of post-war incentives for education, university enrolments doubled between 1944 and 1947, and new fields of applied science, unknown in the 1930s, were emerging. These changes, along with the notion that knowledge had its own economic value, led to the awarding of research grants to universities by the federal government beginning in 1951.

Alan E. Cameron called this phenomenon the "decentralization" of science and engineering, and the rationale for maintaining a strong, central organization. He believed that the CIM had nothing to fear from the establishment of smaller technical units because it could still act as a vital link between disciplines, through meetings and publications. With a large membership reaching into almost every corner of industry, academia and government, the Institute could exert considerable power in joining with other professional institutions in a common cause.[264] Discussions of whether the parent organization was at the risk of dissolving into independent parts had been around since the 1930s, and would crop up again in the 1960s and 1990s. However, if Council had shied away from creating technical Divisions, the CIM might have lost many of its newer members to other organizations as scientists and engineers struggled to specialize in a highly competitive professional environment.

---

[263] In 1948, the Division chairmen were added to Council as *ex-officio* members. *Bulletin*, no. 437 (September 1948): 516.

[264] A. E. Cameron, "The Institute and the Industry," *Transactions*, vol. XLIX (1946): xvii-xxii. The Institute, in conjunction with the EIC, had already been involved in this capacity. It had protested the actions of the Wartime Labour Relations Board, which threatened to include engineers in the collective bargaining process between management and labour, which would result in the engineers being dominated by a trade union of non-professionals. See "Collective Bargaining and the Engineer," *Bulletin*, no. 385 (May 1944): 334-340.

The Jubilee Year

In 1948, the Institute held its Golden Jubilee Annual General Meeting in Vancouver. This was the first time that the AGM had hosted west of Winnipeg. If the twenty-fifth anniversary in 1923 was a modest affair, the fiftieth made up for it. The 1,130 registered delegates included people from "sister" institutions and professional organizations around the world, and there were many distinguished guests. As Charles Camsell passed the presidential honour to Randolph Diamond, many members paused to reflect on CIM's accomplishments, and their great fortune in having been guided by a number of presidents of such outstanding calibre. Membership had risen from eighty-six in 1898 to 3,134 in just fifty years, with 642 affiliated student members. The new technical Divisions were rapidly broadening the Institute's knowledge base, and the Publications Committee had been given permission to increase the volume of technical literature, including the release of a special Jubilee volume entitled Structural Geology of Canadian Ore Deposits. This volume was the first of the Institute's Special Volume Series. Twenty-one Branches, with more on the way, were supporting the industry and bringing the CIM to members in many remote areas across Canada. In 1946-47, President Richard J. Ennis, assisted by 'Curly' Carlyle, had visited most of them by air.

In a short article highlighting CIM's history, W.M. Goodwin reminded members that although:

> many have contributed to the splendid result we see today, it has been achieved primarily by young men, or by older men of youthful spirit. In looking to the future, therefore, and in preparing for it in the present, it behooves us all to encourage our young men to take hold of the industry and of the Institute, and to back them up to the uttermost. Only thus can we ensure that the second half-century will equal the proud record of the first.[265]

Secretary-Treasurer Gerow

On July 1, 1949, Carlyle (Jerry) Gerow was appointed the Institute's Secretary-Treasurer when E.J. Carlyle retired as Secretary Emeritus

[265] W.M. Goodwin, "The Institute: 1898-1948," *Bulletin*, no. 431 (March 1948): 119-121.

after eighteen years of service. To become familiar with the complex duties attached to the position, Gerow would work as an undersecretary to Carlyle until October. Like those of his predecessors, Gerow's qualifications were impressive. Born in 1896, in Bloomfield, Ontario, he first worked as a public school teacher. When World War I broke out, he joined the Canadian Field Artillery, and later, the Royal Air Force. After the war ended, he earned a degree in chemical and metallurgical engineering at Queen's University, and then joined the Dominion Steel and Coal Corporation where he became a specialist in steam-plant engineering. During World War II, he supervised the bunkering of coal-burning convoy ships, then directed heating fuel supplies for all RCAF and RAF training schools in Canada. His next sixteen years as Secretary-Treasurer would span some of the most interesting and eventful times in Institute history.

Post-War Recovery

For those who had weathered the Depression and survived the war, the next two decades were a time of cautious optimism. Careful planning had averted inflation while industrial innovation, foreign investment, Keynesian economics, and government funding had safeguarded a growing economy. Despite these successes, the notion that a recession could return at any time was hard to dispel, and 51% of the population believed that a third world war was not only possible, but imminent.[266] The outbreak of the Korean War in 1950 did nothing to allay their fears. As citizens grew more anxious over the negative aspects of nuclear technology, they also trusted in government aid for education, old-age pensions, baby bonuses, and welfare payments.

Canada had also expanded in population and territory when Newfoundland entered Confederation in 1949. The rise in birth rate, the influx of immigrants, and reasonable veterans' mortgages stimulated the construction industry, while growing consumerism supported the metal and manufacturing sectors. Although there were two brief economic setbacks in 1948-49 and 1953-54, a swift recovery restored public confidence. During a mild recession in 1957-61, many talented Canadians sought work south of the border in what later became known as the "brain drain."

[266] W. Kilbourn, "The 1950s," in *The Canadians 1867-1967*, J.M.S. Careless and R. Craig Brown, eds. (Toronto: Macmillan, 1967) p.328.

H. Mortimer-Lamb was the Institute's second Secretary, from 1904 to 1919. With unflagging zeal and energy, he helped to guide the Institute during the critical years following the death of B.T.A. Bell. During his tenure, the Institute's membership increased from 435 to 1,331, and his staff of one was "handsomely doubled." He single-handedly edited and published the Bulletin and the Transactions. After his retirement, he continued to serve the Institute as Secretary of the British Columbia Division until 1941. (National Archives of Canada photo C144210)

Cartoon published in the Montreal Daily Herald on April 6, 1932, during the 34th Annual General Meeting, depicts six Institute Presidents: F.D. Adams (1910-12); F.W. Gray (1931-32); J.A. Allan (1932-33); M. Dwyer (1937-38); Charles Camsell (1947-48); and A.O. Dufresne (1950-51).

Two metallurgists who helped to build Cominco into one of Canada's most successful mining and metallurgical enterprises. S.G. Blaylock (right) ran the company from 1919 to 1945, and was succeeded by R.W. Diamond (left), from 1945 to 1956. Blaylock was CIM President in 1934-35, and Diamond was CIM President in 1948-49. Both men were inducted into the Canadian Mining Hall of Fame. (Photo courtesy of Cominco)

S.G. Blaylock with a member of the grievance committee in his office at Cominco. Blaylock had a trail-blazing record in both human and industrial relations, and mining and metallurgy. He used to remark: "Three things are necessary to find a mine - brains, guts and luck." (Photo courtesy of Cominco)

Wing-Commander C.H.E. Stewart served as a pilot in both World Wars. A mining engineer and member of CIM, he was on the staff of Anglo Huronian Limited before re-enlisting.

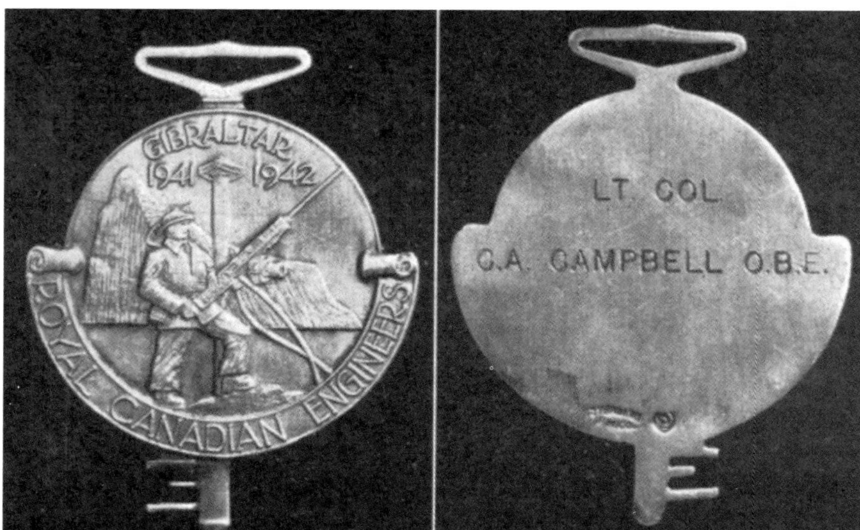

The 'Key to Gibraltar' was awarded to CIM member Lieutenant-Colonel C.A. Campbell and other Tunnellers of the Royal Canadian Engineers in recognition of their work in strengthening the fortifications at Gibraltar. At times they did some 'real' mining. In 1942, Canadian Tunnellers reopened the Carrock mine in the English Lake District.

A.A. Cole (centre), CIM President in 1916-18, having breakfast with L. Belanger (left) and L.H. Cole (right) at Moose River Crossing, northern Ontario, in July 1935. (National Archives of Canada photo PA14901)

A.O. Dufresne (second from left), CIM President in 1950-51, with his son Cyrille Dufresne (third from left), and Marc Boyer (left) at a Labrador Mining and Exploration Company drill site near Knob Lake, Quebec, in 1946. A.O. Dufresne was then Quebec Deputy Minister of Mines, Boyer was Assistant Deputy Minister in the federal Department of Mines and Technical Surveys, and Cyrille Dufresne was a supervising geologist for Labrador Mining. (Photo courtesy of C. Dufresne)

Robert A. Bryce, 1942-44 CIM President, being inducted as Chief O Nee Sta Po Ka (Calf-Child) of the Blood Indians in Alberta in 1950. Bryce was then president of the Canadian Chamber of Commerce. (National Archives of Canada photo C144206)

At the 53rd Annual General Meeting in Quebec City in 1951, from left to right: M.F. Goudge, Vice President, Ontario East; A.O. Dufresne, Past President; Oliver B. Hopkins, President Elect; and D.G. Burchell, Vice President, Maritimes. Burchell became CIM President in 1958-59.

At the 73rd Annual General Meeting in Quebec City in 1971: J.H. Schloen, President Elect; H.M. Wright, Past President; and Tom Patching, 1971-72 President.

Jean-Paul Drolet, 1978-79 President, explaining a point. Drolet was Assistant Deputy Minister (International Minerals) in the Department of Energy, Mines and Resources. He and his wife Françoise conceived and endowed the Order of Sancta Barbara to recognize the contributions of women to the welfare of Canadian mining communities.

(Above) At the opening session of the Tenth Commonwealth Mining and Metallurgical Congress in Ottawa on September 3, 1974: Dr. John Convey, Congress Chairman; W.H. Gilchrist, CIM President 1974-75; Alfred Powis, Congress President; The Honourable Donald S. Macdonald, Minister of Energy, Mines and Resources; 3C.C. Huston, Congress Vice Chairman, and CIM President in 1954-55; Sir Ronald Prain, Chairman of the Council of Commonwealth Mining and Metallurgical Institutions. Convey was then Director of the Mines Branch, and Powis was President of Noranda Mines Limited.

Departure from Edmonton for a Congress tour to Resolute and Ellesmere Island in the high Arctic, with stops in Uranium City and Yellowknife.

Don Worth, Congress Treasurer, and CIM President in 1996-97, and a colleague, en route to Resolute.

Of particular concern to some Canadians was that many of their key industries were controlled by Americans, and that their powerful southern neighbour would soon begin to interfere in domestic and international affairs. American investment hovered at 75% of all foreign investment in Canada, and the United States absorbed over two-thirds of Canadian exports. By the mid-1960s, the proportion of investment would rise to 90% in auto manufacturing, 70% in oil and gas, and 50% in mining and smelting.[267] The growing American presence in the oil and gas industry would soon affect the Institute as the Society of Petroleum Engineers (SPE) of AIME would attempt to establish its own branches in the oil producing regions of Alberta and Saskatchewan, and compete with the CIM for Canadian members.

Canada's post-war industrial growth was characterized by mammoth projects and innovative ideas. The completion of the St. Lawrence Seaway in 1959 brought ocean shipping two thousand miles inland. Iron ore from Labrador and northern Quebec could now be economically shipped to Canadian and American steel plants. In Ontario, the dewatering and dredging of Steep Rock Lake to expose a rich iron ore deposit, one of the most impressive engineering feats in Canadian mining history, was nearing completion. Shortly after the discovery of the Leduc oil field in 1947, construction had begun on the Interprovincial Pipeline.[268] It remained the world's longest, until the completion of the TransCanada Pipeline in 1958, and was viewed by C.D. Howe as a modern version of the Canadian Pacific Railway.

The most sensitive issues affecting U.S.-Canadian relations centred on nuclear technology and strategic defence. Cooperative projects such as the installation of the Distant Early Warning (DEW) Line from Alaska to Baffin Island, and the agreement on a North American Air Defence Treaty (NORAD) were two of the most significant. But as Americans became more protective of their defence industry, Canada's military research and development withered. A prime example was the federal government's cancellation of the Avro Arrow fighter plane in 1959. Instead, Canada decided to build state-of-the-art communications satellites, the first being Alouette I, launched on September 29, 1962. In the nuclear field, Atomic Energy of Canada Limited developed its Candu reactor. The first nuclear energy power plant at Rolphston, Ontario, completed in 1962, became a showpiece of Canadian innovation.

---

[267] Collins, ibid., 195.

[268] McKenzie-Brown, Jaremko, and Finch, ibid., 50.

A Mining Boom

For Canada's mining industry, the post-war years were boom times. In 1947, the Iron Ore Company of Canada (IOC) began the development of the huge iron ore deposits in the Quebec-Labrador Trough. This massive project entailed the building of the 350-mile Quebec, North Shore and Labrador Railway, hydro-electric power plants, and ore-handling facilities to serve the mine at Knob Lake, and the new town of Schefferville. Other developments followed: IOC's Carol Lake mine and the town of Labrador City; Quebec Cartier's railway, Lac Jeannine mines, and the town of Gagnon; and, the Wabush Iron Company's Scully mine and town of Wabush. Elsewhere in the region, the Quebec Iron and Titanium Corporation developed the large Allard Lake ilmenite deposit.

Intense prospecting for uranium uncovered the major deposits at Beaverlodge, Saskatchewan, in 1952, and one year later, at Blind River, Ontario, which resulted in twelve new mines and the new town of Elliot Lake. The Denison mine became the largest producer of uranium in the world.

In 1953, in rapid succession, the Geco copper-zinc-lead deposit was discovered at Manitowadge Lake in Ontario, and the Brunswick No. 6 and No. 12 lead-zinc-silver deposits in the Bathhurst area of New Brunswick. By 1954, the Sherritt Gordon nickel refinery had commenced production at Fort Saskatchewan, Alberta, using the revolutionary ammonia pressure leach process developed by Frank Forward, who was to become CIM President in 1965-66. Copper was in high demand, and in 1955, Quebec's Campbell Chibougamau and Gaspé Copper mines were brought on stream. Next came the discovery of nickel 400 miles north of Winnipeg, at what was to become Thompson, and Inco's Manitoba Division with its smelter and refinery started up in 1961. That same year, zinc was discovered at Mattagami Lake in Quebec, resulting in a new mine and the town of Matagami.

In British Columbia, the first of that province's large porphyry copper and copper-molybdenum open-pit mines went into operation at Bethlehem Copper in 1962. It was later followed by the Granisle (1966), Brenda (1970), Granduc and Island Copper (1971), and Bell, Gibraltar and Lornex (1972) mines. Saskatchewan's first potash mine started up at Esterhazy, Saskatchewan in 1962. Then, in 1966, one of the biggest base-metal mines in the world opened at Kidd Creek, near Timmins, Ontario, leading to an electrolytic zinc plant in 1972, and a copper smelter and refinery in 1981.

Membership

With so many developments in mining, and the prospect that new Branches would soon be added to those already in existence, the Institute's Secretary-Treasurer, Jerry Gerow, actively began to recruit new members. This task reminded Council of an older dilemma. Ever since the first Branches had been formed in 1902, Council had agonized over how to deal with those who were often very involved locally, but did not become national members of the Institute. Some Branches, like the one established in Calgary in 1949, had been fairly insistent that all their members belong to the Institute. However, in the far-flung mining communities, Branches welcomed men who often lacked the credentials to qualify them as full members. For example, it was very difficult for an equipment salesman to become a full member if he had not attended university. Being an associate member in the 1940s and 1950s was sometimes seen as being second-class.[269]

In 1953, Council struggled with the concept of asking all Branch members to join CIM after a certain period of time. But as no consensus was reached on how to amend the By-laws, the matter was postponed for future study. The issue was reopened again in 1962, when Council argued over revamping the "Rules for Guidance of CIM Branches" as "regulations" whereby Branch members would be forced to join the Institute. Again, no resolution could be reached, and the matter was deferred once more.

The First Women Members

In the spring of 1953, the question was raised as to whether women could become members of CIM. Two young women, one a graduate in metallurgical engineering, and the other a graduate in geology and geophysics, had applied for enrolment as junior members, and both had been attending meetings at affiliated student societies.[270] Applications had also been received from two other women asking for full membership. One of these was Viola MacMillan who had been refused admission as an associate member in 1946 because the By-laws stipulated that "an applicant shall be a man."[271] There was no real opposition to the request to admit women members, but before it could happen, By-laws 3, 4, 9 and 22

[269] C. Bruce Ross, interview with author, February 7, 1996, Toronto.
[270] CIM, *Council Minutes*, April 16, 1953, vol. 7, NAC.
[271] E.C. Allingham, Assistant Secretary, CIM, letter to C.C. Huston, July 22, 1954, vol. 7, NAC.

needed to be rewritten as gender neutral. When the proposed changes were put to a general vote of the Institute's membership, more than two-thirds were in favour. In fact, interest in the subject was fairly high because the ballot return was more than 26%.

Women's participation in the Institute, from the beginning, had been in a supporting role: as guests accompanying their husbands to major meetings and on field trips; and later, as members of auxiliary groups like the Women's Association of the Montreal Branch of CIM, and the Women's Association of the Mining Industry of Canada, both established in 1921. In the early years, the AIME was convinced that "the attendance of ladies accompanying members was greatly beneficial,"[272] and this arrangement also suited the Institute. In 1925, the MSNS had passed a motion to welcome spouses, and gradually a ladies' program was integrated into the annual general meetings.[273] Women had also long been active in mining communities, where they organized social events that supported the local Branches, and brightened their families' lives.[274] Some women attended so many annual general meetings with their husbands that they tended to regard themselves as invisible members of the Institute.

Even though women had become eligible for membership in 1954, there wasn't a flood of new applications. At the time, very few women were pursuing careers in applied science and engineering, and even fewer in fields related to the mineral industry. In 1941, women constituted less than 1% of university enrolments in engineering. Their enrolment in earth sciences was still less than 5% in 1976, with figures for engineering and applied sciences considerably lower. It was not until the 1990s, that the number of women in engineering would reach almost 12%.[275]

Because the mining industry was believed to be too dangerous for women, legislation barring them from working underground was first introduced in the 1930s, and remained in effect until the mid-1970s.[276] In most

[272] Wilson, ibid., 52.

[273] Pigot, ibid., 75.

[274] The Women's Association of the Montreal Branch of CIM, *Minutes*, CIM Archives, Montreal. Marjorie Sheridan, "The Women's Association of the Mining Industry of Canada," and Diana V. Tikkanen, "The Greater Vancouver Mining Women's Association (1974)": unpublished histories, and correspondence with author, 1997-1998.

[275] C.W. Stout, "A Degree of Change," *Perspectives* (Winter 1992): 14-19, and "More than Just Numbers: Report on the Canadian Committee on Women in Engineering," (Ottawa: CCWE, 1992).

[276] R. Santi, *Women in Mining: The Progress and the Problems*, Department of Energy Mines and Resources, MR 152 Mineral Policy Series (Ottawa: Supply and Services Canada, 1976). The United Nations International Labour Organization Convention, Underground Work (Women) Convention, No. 45, 1935, was ratified by Canada in 1966.

mines during this period, women's employment was largely confined to providing health and welfare services, or to clerical work. The exception was in Quebec, where female geologists and engineers were given limited access to the rock face.

In accepting applications for membership, the Institute was far more concerned about proper qualifications than addressing gender problems within the industry. Of the four women who applied for membership in 1953, Viola MacMillan's application caused the most stir. To George B. Langford, Head of the Department of Geological Sciences at the University of Toronto, the point was fairly clear:

> I was pleased to note that there are some highly qualified women who have applied for membership. [However] the *Bulletin* specifically states that membership is for properly qualified technical persons. I am well acquainted with Mrs. MacMillan and her activities, and by no stretch of imagination can she qualify under the rules. . . I sincerely hope that the Council will not be swayed by any false sense of chivalry or mawkish sentimentality in dealing with this case.[277]

Although MacMillan was a prospector and mine developer of note, and President of ViolaMac Mines Ltd., and was to become President of the Prospectors and Developers Association of Canada in 1957, she did not have any academic credentials other than public school, which was common for many women born in 1904.[278] She was fortunate in having two credible sponsors who believed that her practical experience compensated for her lack of formal education. C.C. Huston, CIM President in 1954-55, wrote, "She is an active and vocal spokesman for the industry and... it is my considered opinion that few Canadians today can claim to have created and maintained so great an interest in this industry."[279] In a second letter, H.L. Roscoe, Vice President of Noranda Mines Limited, wrote that "Her success in finding a mine and financing it to a very successful conclusion is further evidence of her qualification. . ."[280] MacMil-

[277] G.B. Langford, letter to Council, October 4, 1954, vol. 7, NAC.

[278] J.J. Byrne, "Viola R. MacMillan," *The Discoverers: A 50-Year History of the Prospectors and Developers Association, Some Famous Prospectors and their Discoveries*, Monica R. Hanula, Chief Editor, Roy M. Longo, Associate Editor (Toronto: Pitt Publishing, 1982) p. 251-253.

[279] C.C. Huston, letter to the Secretary, CIM, August 13, 1954, vol. 7, NAC.

[280] H.L. Roscoe, letter to the Secretary, CIM, August 27, 1954, vol. 7, NAC.

lan was granted admission, and became a Life Member of CIM in 1979. She was also inducted into the Canadian Mining Hall of Fame in 1991, and awarded the Order of Canada in 1993.

Pauline Moyd, the other applicant for full membership in 1953, was twenty-three years younger than MacMillan, and had earned a bachelor's degree in geology at Columbia University in 1938, followed by a master's degree at Bryn Mawr College. She had been in a "position of responsibility" for over six years as a consulting geologist in partnership with her husband, Louis Moyd, in New York State.[281]

The two candidates for junior membership also met the Institute's requirements. Harriet I. Chmara was an engineer-in-training at Fleet Manufacturing Ltd., in Fort Erie, Ontario, and Anita G. Evans was a geologist with the Photographic Survey Corporation in Toronto. These women were the first of many who would join the CIM, and in future years, some would be elected to positions of leadership.

Sixth Commonwealth Mining and Metallurgical Congress

From September 8 to October 9, 1957, the CIM hosted the Sixth Commonwealth Mining and Metallurgical Congress, an event which members would recall with pride for many years thereafter. Delegates had travelled to Canada for the second congress in 1927, and then to South Africa for the third in 1930. The fourth congress, planned for Australia in 1933, had to be postponed due to the Depression, and then the outbreak of World War II had pushed the date ahead to 1949, when Great Britain extended its hospitality. The fifth congress took place in Australia and New Zealand in 1953. Formerly known as the Empire Mining and Metallurgical Congress, it changed its name to the Commonwealth Mining and Metallurgical Congress in 1955.[282]

For the Canadian Congress in 1957, the organizing committee decided to place the country's mineral industry on display by means of a comprehensive cross-country excursion. This would give delegates a first hand look at various operations, and provide maximum opportunity for informal discussion. The Congress opened in Vancouver, and the agenda

[281] "Personality Profile, Pauline Moyd," *Bulletin*, no. 777 (January 1977): 50-52.

[282] L.G.R. Crouch, ed., *Proceedings, Sixth Commonwealth Mining and Metallurgical Congress, Canada, 1957* (Montreal: General Committee of the Sixth Commonwealth Mining and Metallurgical Congress, 1957).

included travel by railway from Vancouver to Sydney, with side trips by air to northern British Columbia, the Yukon, the Northwest Territories, and parts of northern Saskatchewan, Manitoba, Quebec, and Newfoundland. Whenever the entourage reached a major centre, meetings were arranged by the local CIM Branches, and the delegates visited nearby mining, metallurgical, and oil and gas operations, and other points of interest.

In total, 479 people from thirty-five countries participated in the Congress. Technical meetings took place in thirty centres, and more than fifty mines and fifty metallurgical and other plants were visited by individual groups. The five special Congress Volumes prepared by CIM's technical Divisions were also a great success, and became standard reference texts. They included *Geology of Canadian Ore Deposits*, *The Geology of Canadian Industrial Mineral Deposits*, *Case Histories in Geophysics*, *Mining in Canada*, and *Milling of Canadian Ores*. An extremely fine account of the Congress, containing speeches, photographs, and anecdotes, was compiled and written by former CIM President George E. Cole, assisted by his wife. Unfortunately, Cole would pass away before the final version was finished, but its production was faithfully carried out by Mrs. Cole and other Institute members. It remains not only a valuable record of this important Congress, but also a unique window on Canada's mineral industry in 1957.

The Education Dilemma

In his welcoming address at the Sixth Commonwealth Congress, President Randolph W. Diamond spoke at length about Canada's need for well-trained engineers, geologists, and metallurgists. He predicted that the serious shortage of qualified people would soon hinder progress within the mineral industry, and the country would begin to lose its edge in science and technology. Why was this happening when university enrolments were climbing? Among the reasons offered, Diamond claimed that too many students were leaving prematurely in favour of industrial jobs, and large numbers of graduates were finding better opportunities south of the border. Even more disturbing was that students seemed to be more interested in such exciting new fields as aeronautics and electronics, rather than in "mining [which] has lost much of its glamour in the eyes of our young men [and] the same seems to be true of metallurgy."[283] His answer was to stimulate public awareness in the

[283] R.W. Diamond, address to delegates, Crouch, ibid.

importance of the "non-ferrous metal business," starting from grade school. His plea was to be taken up again in the future.

The Institute's interest in education dated back to 1918 when it formed the first Committee on Education. Since then, it had maintained a watch over educational matters affecting the industry. In 1956, because of the declining number of qualified graduates, the Committee on Education for the Mineral Industry organized a special forum to promote discussion between educators and industrialists. There were several surprising conclusions. The shortage in engineers was not confined to the mineral industry or to Canada; the industry often used engineers to do the work of technicians, and university administrators lacked a basic plan for handling any increase in students who wished to follow science or engineering programs.

There was also debate over the government's new policy of issuing research grants to universities, a practice initiated in the early 1950s. Horace J. Fraser, a future CIM President (1957-58) who had taught for seven years at the California Institute of Technology, felt that schools might forfeit their academic freedom if they relied too much on federal assistance. Instead, he called on "production" to provide funds for scholarly work. Thirty years later, the concern would be exactly the opposite: that financial assistance from industry would lead to a loss of academic freedom.

Six years later, the predictions made by Diamond were proving to be correct. During his Presidential address at the Annual General Meeting on April 25, 1962, James B. Mawdsley emphasized that "the number of young trained engineers and scientists entering the Mineral Industry is shrinking and is far below present and future needs."[284] Mawdsley was well-placed to comment on the situation. He had taught geology for more than thirty years at the University of Saskatchewan, and had recently been appointed Dean of Engineering. After examining the statistics on science and engineering students, he found that mineral industry careers remained their least favourite choice. Of 2,500 engineering graduates in Canada, only 200 had specialized in either mining, metallurgy, geology, or petroleum engineering. Honours geology and geophysics added another eighty. These numbers were totally inadequate to meet the needs of a growing and complex mineral industry. Mawdsley's conclusions were very similar to those of Diamond and Fraser: it was vital that industry invest in its future by providing as many scholarships and grants as possi-

[284] J.B. Mawdsley, "The Mineral Industry Needs More Trained Men," Presidential Address, CIM Annual General Meeting, April 25, 1962, vol. 7, NAC.

ble, and that everyone in a position of authority within the mineral industry act as its ambassador.

One man who would fit this description perfectly was John R. Bradfield, CIM President in 1962-63, and a man who had spent his entire career at Noranda, becoming its Chairman and Chief Executive Officer in 1964. Bradfield had a life-long interest in education, and his advice to young people was "to stay in school long enough to complete their education and fully develop their natural talents."[285] Along with stimulating student interest in the mineral industry, he propped up the ailing mining departments at McGill and Queen's Universities in the 1950s, and created and chaired the Canadian Mineral Industry Education Foundation.[286] At Noranda, he was known for allowing his technical and scientific staff the opportunity to explore new methods and procedures, which proved to be of tremendous benefit to individuals as well as the company. Bradfield was oficially recognized for these achievements in 1973, when he was awarded both the Institute's Inco Medal, and the Order of Canada.[287]

In 1964, the General Committee on Education became the Advisory Committee on Education. It helped foster communications between academics and members of the Institute's technical Divisions, many of which had already formed their own sub-committees on education. Some Branches, like the one in Hamilton, had established a student loan fund, and others offered scholarships. For younger students, special projects had already been developed to encourage careers in engineering, and the Institute and its Branches had supported science fairs at local elementary and high schools. CIM had also produced and distributed a career brochure, as well as a film called "Nation Builders," and later, designed a slide show for use as a teaching tool in the classroom. Another booklet entitled "Careers in the Mineral Industry" was so successful that several provincial governments ordered large quantities for their schools, and the Institute had arranged to have a French translation available for Quebec.[288] The seeds for success had been sown, but germination would take time and patience.

---

[285] "Emphasis is on Education In Building Mining Giant: Noranda's John Bradfield," *Northern Miner*, November 28, 1968.

[286] This is not to be confused with the Institute's Canadian Mining, Metallurgical, and Petroleum Foundation.

[287] "To J.R. Bradfield: the Medal of the International Nickel Company of Canada," *CIM Directory*, vol. 7 (June 1973): 77.

[288] CIM, *Council Minutes*, November 23, 1962, vol. 7, NAC.

Fifty-Year Club

In 1958, Council marked the Institute's sixtieth anniversary by introducing the Fifty-Year Club, a mark of respect and appreciation for those who had served the Institute for an aggregate period of fifty years. Inductees were awarded a special certificate and lapel pin which were presented by the President at the Institute's Annual Dinner. When members reached the magic number, they received a congratulatory letter, usually written by the Institute's Secretary or Executive Director.

When H.E.T. Haultain was informed that he had become a member, he wrote to Jerry Gerow: "I see in copy of the *Bulletin* an account of the inauguration of the fifty-year club. I am puzzled. Why a club – has it any officers, or duties, or any dues, which I think are general characteristics of a club?"[289] Haultain had always been extremely cautious about accepting awards, and had once refused the designation of Honourary Member even though the nomination was supported by over 900 ballots. E.J. Carlyle, who addressed Haultain as "dear old bear," begged him to reconsider as a show of respect for his Institute colleagues, but Carlyle was unable to persuade him and the nomination was dropped. What was important to Haultain was that he not receive anything which he considered beyond his due. Having already accepted Life Member status in September 1943 at the age of seventy-four, his poke at the Fifty-Year Club was probably for amusement.

A Serious Deficit

During the 1960s, the Institute had fallen into a serious financial deficit position, and Divisional requests for more publication space were adding to the burden. With the addition of the Mechanical-Electrical Division in 1961, page allotments in the *Bulletin* were causing many arguments. Council's response to the financial crisis was to raise membership dues. As many members "had no idea of the serious condition of the Institute finances," Branch secretaries were asked to inform their constituencies.[290] A ballot was mailed to all members during the summer, and the new dues structure was approved. It went into effect on January 1, 1962. Although the lack of funds was a continuous worry, it was not the only concern. During this era, Council was also embroiled in other major issues: the publication

[289] H.E.T. Haultain, letter to C. Gerow, July 11, 1958, vol. 7, NAC

[290] CIM, *Council Minutes*, Proposal to Amend the By-laws, February 17, 1961, p. 1958, vol. 7, NAC.

problem, the western field office, a proposed change of name, the reorganization of Council, and the possible relocation of Headquarters. To make matters even more complicated, the Executive Directorship was in transition.

A Change in Helm at Headquarters

On August 1, 1963, Richard E. (Dick) Barrett was appointed Executive Director after Jerry Gerow retired as Executive Director Emeritus. Like the previous Executive Directors, Barrett was well-known and well-qualified. He had earned a BSc in mining engineering at McGill, then after working for several years in the coal and gold mining industries, he taught at the University of Toronto, becoming Head of the Mining Engineering Department in 1948. Five years later, he joined Eldorado Mining and Refining Limited as Manager at Beaverlodge, then moved to Ottawa in 1955 to become its Director of Ore Production.[291] Having been a CIM member since 1934, a Chairman of the Ottawa Branch, and a District Councillor, he was very familiar with the Institute. However, after only two years as Executive Director, and at the age of sixty, he chose to retire on October 1, 1965.

Barrett's successor was E. Gordon Tapp. His appointment as Executive Director was unique in that he was promoted from the ranks of the Headquarters staff. He had been appointed Assistant Secretary of CIM in 1954, and had then edited the *Bulletin* before being promoted to Secretary-Treasurer in 1963. Along the way, he had earned a reputation as a hard-working and compassionate man who got things done. Tapp's former position as Secretary-Treasurer would be filled in 1965 by Gordon Skilling, another key member of the staff who would eventually replace Tapp in 1979.

Gordon Tapp had graduated from Queen's University with a BSc in mining engineering, and had then joined the Consolidated Mining and Smelting Company (Cominco) at Sullivan, B.C., in 1941. He had worked in various jobs from underground surveyor, to testing engineer, and then assistant supervisor. In 1948, he transferred to the Kimberley operation as a cost engineer for the company's mine, concentrator, and fertilizer plant. Along the way, he had become a CIM member, and a founding member and first Chairman of the Sullivan Branch. Tapp would stay at CIM for the next fourteen years, and like those who came before him, he would leave his imprint on the Institute.

[291] Barrett was succeeded as Manager at Beaverlodge by W.M. Gilchrist, CIM President in 1974-75.

Publications and Divisional Unrest

In 1960, the controversy over publication page quotas and the serious backlog of unpublished papers would cause so much anxiety that one division would talk openly of leaving the CIM. At a Council meeting in April 1960, H.T. Airey, Chairman of the Metallurgy Division, read a resolution which had been approved a few days earlier by his division:

> The Metallurgy Division of the CIM wishes Council to be informed that the Division deplores, in no uncertain terms, the present situation regarding publication of technical papers, in the *Bulletin* and *Transactions* of the Institute, and recommends that the system be drastically revised.[292]

Although the situation did not affect them equally, most of the Divisions had voiced alarm over the few papers that had been published, while realizing that adding pages to the *Bulletin* would probably decrease advertising revenue and add to the Institute's deficit. However, the Geology Division cautioned that "the circulation of the *Bulletin* as well as its usefulness to the Institute and to the Canadian mining industry are, and will always be, a function of the number of pages published."[293] Its interim solution was to ask its authors to shorten their papers so that more could be accepted. The Petroleum and Natural Gas Division's answer was already confirmed: it would seek its own journal as soon as possible. The *Journal of Canadian Petroleum Technology (JCPT)* was launched in 1962.

For the Metallurgy Division, the publication issue was closely tied to an internal reorganization. Opposition at Headquarters to printing a satisfactory number of metallurgical papers could have meant a lack of support in other areas:

> A separate Metallurgy Division publication has been put forward by the Committee as a means of strengthening the technical status of the Metallurgy Division. . . if some reform is not adopted in the near future, the Committee believes that the formation of an independent metallurgical society in Canada will be inevitable.[294]

[292] CIM, *Council Minutes*, April 28, 1960, p. 1034, vol. 7, NAC.
[293] CIM, *Council Minutes*, March 23, 1961, p. 1117, vol. 7, NAC.
[294] Ibid., 1133.

The impasse was breached in 1962 after John Convey offered the resources of the Mines Branch for the production of a *Canadian Metallurgical Quarterly*. As for another concern of the Metallurgy Division, the Institute would announce in 1963-64 that representation for technical Divisions on Council would increase to two members each.

The question remained whether creating independent publications was good for the Institute. The *Bulletin* had always been an integral component of membership, and without a well-rounded publication, CIM's reputation could be diminished. As the new *Journal of Canadian Petroleum Technology*, and the *Canadian Metallurgical Quarterly* gained in acceptance, and a geological quarterly was being considered, many feared that the *Bulletin* would lose its impact. Even more serious was the concern that if the Divisions became too powerful, they might eventually seek total autonomy.

The Rise and Fall of the Western Field Office

In 1958, a special committee was formed to study the feasibility of a CIM office in Calgary. Two years later, John C. (Cam) Sproule, then CIM President (1959-60), announced that the committee had considered investing between $20,000 and $25,000 for a Calgary office over a two-year period, and that if successful, it would become a permanent fixture. The timing was certainly good. By 1960, the Petroleum and Natural Gas Division was the fastest growing group within CIM. Not only had it opened its own technical library in 1955, supported by donations from the Alberta Society of Petroleum Geologists, the Oilfields Technical Society, as well as the CIM, but starting in 1960, the Division was the first to hold separate annual technical meetings lasting several days. In anticipation of future needs, Sproule had also insisted that cooperation be sought with AIME through Stan Harding, then Chairman of the Petroleum and Natural Gas Division.

The Western Field Office received Council approval on October 27, 1961. It was organized to serve Districts 4, 5, and 6, create stronger ties with Headquarters, promote membership in the Petroleum and Natural Gas Division, and develop a special quarterly publication for that division. The logical candidate for western field secretary was John G. Ditchburn, future editor of the *Journal of Canadian Petroleum Technology*. The office was rented from Sproule Projects in Calgary, and Ditchburn hired a stenographer to assist with editorial duties.[295] When the 1963-64 CIM President, William J. Johnson, toured Calgary and Edmonton with Richard Barrett in 1963, the progress report on the Western Field Office was very promising, but the two men felt that the office needed a more central location. As for the JCPT, it was rapidly becoming a highly respected publication within the oil and gas industry, and was projected to be financially self-supporting within two years.

For reasons which are not entirely clear, Council decided to close the Western Field Office in December 1964 on a recommendation from the CIM-AIME Committee of the Petroleum and Natural Gas Division.[296] Wayne Glen, the Committee's chairman, advised CIM's President, Ralph D. Parker, that there were not enough funds to both operate the field office and publish the *JCPT*.[297] Although the poor financial state of the Institute was a factor, other pressures may have included the AIME's Society of Petroleum Engineers goals for expansion into the Canadian 'oil patch.' Ditchburn continued in his post until November 30, 1965, but the office space was vacated, and the business operations, including the editing and publication of the *JCPT*, were transferred to Montreal. It is hard to imagine how this turn of events affected morale within the Division, but one year later, it applied for new status as the Petroleum Society of CIM. It would take another thirty years before the *JCPT* would return to Calgary.

A Library for Sale

The financial difficulties of 1964 forced Council to rethink the importance of the Institute's library. As old as the Institute itself, much of the collection was as rare as it was historic. The holdings now contained several thousand items, most of which had been catalogued and indexed. Many volumes charted the early techniques of mining, metallurgy, geology, and related fields, and documented the activities of CIM "sister" societies world-wide.[298] The library was also a repository of everything that the Institute had published since 1898. Unfortunately, it could not compete with the much larger library facilities of universities, and government institutions such as the Mines Branch and the Geological Survey of Canada, nor could

[295] CIM, *Council Minutes*, October 27, 1961, vol. 7, NAC. "Western Field Office," *Bulletin*, no. 609 (March 1963): 178.
[296] CIM, *Executive Committee Minutes*, Minute 47, November 13, 1964, vol. 7, NAC.
[297] "Report of Council, C.I.M.: The Western Field Office," *Transactions*, vol. LXVIII (1965): 139.
[298] Among the library's treasures was an original 1561 edition of Agricola's classic *De Re Metallica*. This book was not included in the collection sold to Laurentian University.

it keep pace with the information explosion. It was becoming less relevant to the CIM membership. It also required the services of a full-time librarian, and its acquisitions were putting pressure on the limited space. Sentiment over the library, and tradition, would lose out to reality.

On October 29, 1964, the Library Committee submitted two recommendations to Council. The Institute library should be abolished because not enough members were using it, and it should be sold 'en bloc' to any Canadian university that tendered a reasonable offer. Efforts were made to appraise the collection's value, and letters were sent to all Canadian universities inviting bids and advising that the price would be negotiable. Of four bids tendered, the highest was received from the newly established Laurentian University. In August 1965, Gordon Tapp, assisted by the university's representatives, crated and shipped the collection to Sudbury.[299] Not everything was sold, however, as the Institute kept its complete set of in-house publications, including the rare *Transactions* and *Proceeding* from the turn of the century.

Reorganization of the Institute

By the mid-fifties, CIM's membership had slowly climbed to an all-time record of 5,000. The task of serving so many members was putting enormous pressures on the Institute and its resources. With thirty-seven Branches in existence, it was no longer possible for the President to attempt the extensive tours made by his predecessors, and in 1956, Alfred L. Penhale asked the five regional Vice Presidents to take on the responsibility of visiting Branches within their districts. This would allow the President more time to attend to Institute business and his many other duties. As for Headquarters, it had outgrown its rental space at the Drummond Building long ago, and new prospects were once again being considered.

By 1962, the situation had not improved, costs had risen, and the Institute was in financial difficulty. This prompted the formation of an Ad Hoc Committee consisting of James B. Mawdsley, President, John Bradfield, President Elect, and four others to solicit ideas from the Divisions and many of the Branches. All aspects of the Institute were scrutinized, such as its organization, publications, possible regrouping of Divisions, and the relation of Headquarters to the Divisions, Districts, and Branches. Affilia-

[299] E. Gordon Tapp, interview with author, July 7, 1996, Belleville, Ontario, and "Institute Library Sold to Laurentian University," *Bulletin*, no. 640 (August 1965): 858.

tion with other societies was also under discussion, as was the Institute's relationship with AIME. The Committee concluded that "Undoubtedly, everyone will not be satisfied but it seems likely that there will be changes made which should strengthen [the] Institute."[300]

On November 23rd, under the presidency of J.R. Bradfield (1962-63), the By-laws were amended to reflect a change in the composition of Council whereby District representation would drop to four members for each, and that of each of the seven technical Divisions would increase to two, namely the Chairman and Vice Chairman. This arrangement, which would begin in 1963-64, came about largely because the Divisions were asking for more representation on Council. The total number of Council members would then be forty-eight, the highest since the Institute's founding.

In its final report, the Ad Hoc Committee also proposed two other major changes:

> As recommended in the report of the Committee, and after discussion, it was moved and seconded that the Council recommend to the membership that the name of the Institute be changed to The Canadian Institute of Mining, Metallurgy, and Petroleum [and]... that Council take the necessary steps to initiate the action required to move Headquarters from Montreal to Toronto.[301]

These initiatives would need approval from the membership. However, a new name could only be realized through an Act of Parliament, and leaving Montreal would also necessitate a change in the Institute's Charter.

Proposed Name Change

The idea for a new Institute name was first discussed in 1960 when members of the Petroleum and Natural Gas Division, the Calgary Petroleum Section of CIM-AIME, and the Calgary Branch, urged Council that the existing name was not representative of the breadth of the technical Divisions, and could be a hindrance to recruiting new members. The search was on for a name that could adequately describe the Institute's activities. Suggestions included the Canadian Institute of Mineral Resources, the

[300] CIM, *Council Minutes*, April 26, 1962, p 1222, vol. 7, NAC.
[301] CIM, *Council Minutes*, November 23, 1962, minute 22, vol. 7, NAC.

Canadian Institute of Mines and Minerals, and simply the Canadian Institute of Minerals, but the choice of Mawdsley's Ad Hoc Committee remained as the Canadian Institute of Mining, Metallurgy and Petroleum.

Historically, name changes have always presented a challenge for multi-disciplinary groups. Like the addition of the word "metallurgy" to CIM's name in 1920, the 1960 proposal was hotly debated. A similar experience had happened to the South African Institute of Mining and Metallurgy (SAIMM) in its quest for a new name in 1902:

> If we are to introduce mining proper into our deliberations, we must include a number of subjects which are far removed from the consideration of chemists and metallurgists. We have no concern with timbering, winding, drills, dykes, faults, or any of the thousand and one problems connected with mining, which come within the province of the engineer, surveyor, or geologist.[302]

Although SAIMM's stumbling point had been the word "mining", CIM's problem focussed on the word "petroleum." As it stood, the name Canadian Institute of Mining and Metallurgy did not refer to geology and other disciplines either. Critics of the proposed new name stated that while it might be advantageous to define all fields covered by the Institute, the word petroleum was redundant because it was included in the term "minerals." One member thought that besides being clumsy and rhythmically incorrect, the semantics were wrong: "mining and metallurgy are activities [while] petroleum is a product."[303] If the word petroleum did not fit, then perhaps a more simple name would suffice, such as the Institute of Mineral Industries. However, this would also necessitate a change in the CIM abbreviation, recognized worldwide.

As one year followed another, it became clear that despite Council's best intentions, unanimity would not be easy to obtain. For some, a more serious question was not how to label the Institute, but how to safeguard its history: "If we choose an entirely new name, we are not only abandoning a name and tradition of which we should be proud but, also, all the goodwill and recognition that has been built up under that name."[304]

[302] "Early History of the SA Institute of Mining and Metallurgy," *Journal of the South African Institute of Mining and Metallurgy*, vol. 94, no. 7 (July 1994).
[303] W.A. Hesse, letter to the Editor, *Bulletin*, no. 623 (March 1964): 236.
[304] Ibid.

In a letter to Cam Sproule (CIM President in 1959-60) in July 1964, Barrett expressed the on-going need to build a strong Petroleum and Natural Gas Division due to the "lack of cohesion and common interests among the various petroleum groups." After all, "Even the AIME did not change its name to include Petroleum until [that] group numbered in thousands."[305] In his reply to Barrett, Sproule wrote "I can't say how disappointed I am that the CIM have in effect thrown the Oil Industry overboard. Don't they know yet that the capital investment in the Canadian oil industry has been greater over several of the past few years than all other mining industries combined?"[306]

There was very little mention of the name change after 1964. As in most unresolved matters, the issue was destined to return just as the Institute was striving to satisfy new demands from its technical Divisions. The name that would finally be chosen would be identical to the one proposed in 1962, but it would take another thirty years to implement.

A Proposed Move to Toronto

The proposal by the Ad Hoc Committee to relocate the Institute Headquarters to Toronto was one of the most divisive issues in the Institute's history. A move to Toronto had been proposed once before, in 1930, but no action had been taken.[307] Now, a relocation looked promising on paper. The cost of moving the offices was estimated at $20,000. Modern, air-conditioned offices could be rented in several new towers between King, Bay, Queen, and Yonge streets, for about $4.50 per square foot, which was slightly cheaper than comparable rents in Montreal.

For some people, the choice of Toronto had actually made a lot of sense. Since 1960, it had become the fastest-growing city in North America. A new expressway was planned, and the city was developing into an excellent place for television, radio, and publishing.[308] Not only were the head offices of the Mining Association of Canada (MAC) and the Ontario Mining Association (OMA) located there, but if they and CIM could arrange housing in the same offices, then all three could share a receptionist and pool their secretarial needs. In addition, the head offices of Noranda, Inco, and Falconbridge and many other mining companies were in Toronto, and their proximity would be an advantage.

[305] R.E. Barrett, letter to J.C. Sproule, July 3, 1964, vol. 7, NAC.
[306] J.C. Sproule, letter to R.E. Barrett, June 23, 1964, vol. 7, NAC.
[307] *Bulletin*, no. 217 (May 1930): 588-589.
[308] Kilbourn, ibid., 322.

Despite the obvious benefits, a major obstacle would be a legal one. Article 5 of the Institute's Charter stipulated that changing the location of the Institute's Headquarters required the approval of "two-thirds of the members of the Institute." When Barrett consulted the office of the Secretary of State, he was informed that the phrase "two-thirds of the members" was ambiguous and could not be used as a procedural guide.[309] It was noted by the Executive Committee that in the past, no vote had ever recorded ballots from more than two-thirds of the total membership. Some members thought that the phrase could be interpreted as two-thirds of the votes cast, while another member mused that perhaps it had been written that way to make it difficult to move the Headquarters. After legal consultations, and a second visit to the Secretary of State, it was determined that an amendment of Article 5 would remove the ambiguity.[310] In June 1965, a private bill was passed in the federal Parliament to change the Article to read that "The Head Office.....shall be in Montreal....or at such other place within Canada, as the Institute may, by by-law, determine."[311] According to the relevant by-law, a change in location would now need an approval rate of only two-thirds of the votes cast.

With the ambiguities removed from the Charter, it was now possible to start the relocation proceedings. A letter indicating that a new by-law would be written to allow the "Head Office of the Institute [to] be moved to Toronto on or about April 1966" was distributed to the membership.[312] In the meantime, the CIM office lease in the Drummond Building was extended until May 31, 1966, in order to take a possible move into consideration, and not to interfere with the Annual General Meeting in Quebec City. What happened next took place largely behind the scenes, but a general vote never materialized. Those who would have voted "no" believed that Toronto already had too much power, and such a decision would have ultimately split the Institute. Others believed that Ontario's position as the nation's mining centre was a temporary phenomenon, and the next wave of mining industry expansion would happen elsewhere. Many members were also reluctant to forfeit Montreal, the birthplace of the Institute.

On October 21, 1965, it was moved and seconded that Council accept the recommendation of the Executive Committee to keep the Headquarters in Montreal, and Gordon Tapp, the newly installed Executive

[309] CIM, *Executive Committee Minutes*, January 24, 1964, p. 736, vol. 7, NAC.
[310] CIM, *Executive Committee Minutes*, September 23, 1964, p. 769, vol. 7, NAC.
[311] "Amendment to C.I.M. Constitution," *Bulletin*, no. 639 (July 1965): 712.
[312] CIM, *Executive Committee Minutes*, ibid., minute 23.

Director, was asked to secure a new five-year lease in the Drummond Building. The issue of relocating the Institute offices, which would be raised again in future years, and the wrangling over a proposed new Institute name had encouraged an introspection which would not have happened otherwise. In the process, the Institute had matured.

With so many younger, university-trained people taking an interest in the Institute, its leadership was beginning to reflect the needs of a different generation, one which had not been exposed to warfare or economic hardship. Despite continued anxiety over the Cold War, there was hope that new technologies would help Canada's mineral industry to grow and prosper in an increasingly competitive world.

THE MODERN ERA: 1967-1998

The decision to keep the Institute Headquarters in Montreal was a watershed that helped the membership accept who they were, and where they were headed. While its location had not changed, the CIM was certainly different from the one which had existed in the postwar era. Gone was the treasured library, and the *Bulletin* would be redesigned to allow for a new publication called the *CIM Directory*. Even the term Headquarters was being used less frequently, having been replaced by Secretariat.[313] By the late 1970s, CIM would also make greater efforts towards bilingualism in its publications and services. With new representation on Council, the technical Divisions had a stronger voice in the affairs of the Institute, which was becoming even more technically oriented. Concerns over public relations and the environment led to new committees and new activities. There would be some turbulent years ahead, particularly in the area of finances, as the mineral industry and Canadian society would be affected by a recession, the restructuring and downsizing of businesses, and the information age.

Although Canada had progressed to a state of post-industrialization by the 1960s, many companies faced constant adaptation to change. Global communications and international markets began to influence domestic affairs, and new technologies were transforming the workplace. Competitiveness had become a necessity for survival. Western Europe, having rebuilt its infrastructure since the war, was now recording higher growth rates than either Canada or the United States. Ten years later, Japan's industrial sector was out-performing North America's, and the "Four Tigers," Hong Kong, South Korea, Taiwan, and Singapore were vying for their share of the action. To complicate matters further, many industrialized nations were dependent on Middle Eastern oil, and in 1973, the Organization of Petroleum Exporting Countries (OPEC) began to exert its political will. This led to a world-wide oil crisis which lasted until the 1980s.

While Canadians celebrated their Centennial in 1967, government spending peaked on social programs and work incentives. The 1965 Auto Pact, which allowed automobiles and parts to cross the United States border without tariffs, provided a safety net for Canada's automotive industry, and the commissioning of large-scale projects like James Bay Hydro in 1971, led many citizens to believe in their continued prosperity. It was not to last very long, as in 1975, Canada began to slip backwards into a state of

[313] The term "Secretariat" first appears in the 1962 *Council Minutes*.

high unemployment and inflation, and the government introduced wage and price controls to avoid the inevitable recession. Five years later, a controversial National Energy Program (NEP) was created to stimulate oil production in western Canada, increase domestic ownership, and provide the government with a new source of revenue. In 1989, the North American Free Trade Agreement began to take its toll on Canada's manufacturing sector. By 1990, unemployment had reached 11%, and many jobs were disappearing through company restructuring and downsizing. Because many of Canada's industries were dependent on the American economy, events such as the collapse of the U.S. auto industry affected Canadian exports.

Decline and Revival of the Coal Industry

In the 1950s, coal mining, once the backbone of Canada's mineral industry, experienced a devastating set-back with the advent of diesel locomotives, and a shift to oil and gas for domestic and industrial heating. As coal consumption rapidly declined, many mines faced closure. Some miners believed that the situation was temporary, as it was rumoured that a new coal-burning turbine would soon exceed the efficiency of the diesel engine. Unfortunately, their hopes were dashed, and collieries were forced to reduce their workforce in B.C.'s Fernie-Sparwood area, and in Nova Scotia's Sydney Coal field. Then, in 1960, there was a dramatic shift. The growing Japanese steel industry required massive amounts of coking coal in order to supply steel to Japan's shipbuilding, construction and manufacturing sectors. In 1968, a major deal had been signed between Kaiser Coal Limited and Mitsubishi Shoji Kaisha, which represented seven major Japanese steel companies; the contract called for a minimum of forty-five million tons of coal over a fifteen-year period.[314] Other contracts followed. This revitalized the Elk River basin, and other previously neglected fields like the Fording River. In just one decade, Canada's coal production would rise from 86 million tons in 1970, to 932 million tons by 1980.[315]

In effect, the Japanese economy had saved the Canadian coal industry. Although western coal mining experienced some instability from 1972 to 1975, other factors came into play. The oil crisis saw a resumption in the use of certain grades of coal, and as world demand for

[314] Bruce Ramsey, *The Elk River Valley: 100 Years of Coal Mining*, A Regional Community Project, 1898-1998 (Sparwood, B.C.: Ramsey Publications, 1997) p. 234.

[315] "Fuel Production in Canada," Statistics Canada, *The Canadian Global Almanac, 1998*, John R. Colombo, ed. (Toronto: Macmillan Canada, 1998) p. 219.

steel increased, there was a shortage of coking coal. Fording extended its market to Taiwan, and then to Brazil, Chile, and Europe, and also supplied blended coal to Ontario Hydro, Algoma Steel, and the eastern United States.[316]

Throughout the difficult period for the coal industry, the leaders of CIM's Coal Division remained optimistic, even though attendance was very low at the annual meetings. In 1958, they undertook a cooperative study with the Metallurgy Division's Iron and Steel Committee into new uses for coal in iron and steel production. Their perseverance in maintaining the Coal Division in the sixties, and bringing it back to "standing room only" in the nineties can be attributed to their willingness to engage in inter-divisional activities, and to their unshakable faith that coal still had an important future. Men like David G. Burchell, CIM President in 1958-59, Ralph D. Parker, CIM President in 1964-65, and Alexis Ignatieff of the Mines Branch, deserve special mention for having kept the Division's *esprit de corps* alive.

Meetings Committee

The coal saga helped the Institute to realize that it needed to provide more support to the engineers, scientists and technicians who worked in the mineral industry. This meant more publications, more regional and district meetings, more educational courses, and more technical committees. In 1968, a special Meetings Committee was formed to study, plan, guide, and coordinate improvements for all CIM meetings. Drawn from all sectors of the Institute's membership, it included people from finance, publishing, and public relations, and representatives from each sector of the mineral industry, along with their suppliers, and interested university and government departments. Discussions were wide-ranging, but one of the Committee's concerns was the need to keep up to date on new technologies. The Committee recommended that a special plenary session on information processing and computer applications be held at a future Annual General Meeting, and advised that a lead-time of two to three years would be necessary to achieve this. Special plenary sessions were to become a regular feature of the AGM.

---

[316] Mark Ladouceur and Leah Gardiner, eds., *The History of Fording Coal Limited* (Calgary: Fording Coal Ltd., 1991) p. 33-36.

Canada's Centennial Year

Canada's Centennial in 1967 was marked by the Institute by adding the Centennial maple-leaf logo to its letterheads and publications. Although CIM had been invited to participate in the Canadian pavilion at Montreal's Expo '67 World Fair, it declined because the financial commitment was deemed to be too large. However, an outdoor geological exhibit was assembled by groups associated with the Institute. In a presidential address in 1967, Edward D. Brown spoke of the Institute's accomplishments with pride:

> We have been busy this year looking back over 100 years of Canadian history and progress. We have liked what we have seen, we are proud of the advances in the minerals industry since Confederation... God grant us the wisdom to intelligently handle the growing importance of our task.[317]

The year was a good one for the Institute. The Metallurgy Division was renamed the Metallurgical Society of CIM, following the example of the Petroleum Society which had been formed two years earlier. There were now seven technical Divisions and Societies, and the Mining Society of Nova Scotia and forty-nine CIM Branches represented the Institute in Canada's ten provinces. By 1967, the Institute's total membership had surpassed the 7,000 mark.

Order of Sancta Barbara

During Canada's centennial year, a special award was inaugurated to recognize the outstanding contributions of women to the welfare and development of mining communities. The idea had come to Jean-Paul Drolet, a future CIM President (1978-79), when he realized that with a few rare exceptions, the Institute's distinctions were all presented to men. In order to remedy this situation, he and Françoise Desrochers-Drolet created and endowed an annual award to honour the often neglected achievements of the "wives and daughters of miners, who accompanied [the men] in their wanderings through the mining regions of Canada."[318] The *Order of Merit of Sancta Barbara* is named after Saint Barbara, who was adopted centuries ago as the patron saint and protector of miners. The silver medal was designed by Françoise Desrochers-Drolet, a well-known Canadian enamel

[317] E.D. Brown, *Bulletin*, no. 964 (October 1992): 19.

[318] Jean-Paul Drolet, *Sancta Barbara, Patron Saint of Miners* (Hull, Quebec: Published Privately, 1990).

artist, in cooperation with the French medallist, Élie Pellegrin, and depicts Saint Barbara and a mine headframe.

A New Directory and a New Look for the Bulletin

For many years, the *Bulletin* had included a list of the names and addresses of all CIM members in its December issue. With a soaring membership, this was becoming an almost unmanageable task for the Secretariat staff. On a recommendation by Gordon Skilling, the Secretary-Treasurer, a new publication called the *CIM Directory* was introduced in 1967. The cover of the *Bulletin* was redesigned in 1968, and the following year, its date of issue was moved ahead to the first of each month. Within two years, both the *Directory* and the *Bulletin* showed a modest excess of revenues over expenditures.[319] This was good news, especially since the Institute's last few years had been fairly lean financially.

A Loss of Valued Members

When James M. Harrison took office as President in 1969-70, he had no idea that the Institute was about to lose three of its most influential members. In November, a jet carrying six key mining executives crashed in bad weather on a flight from Churchill Falls to Labrador City. Among the dead was Donald P. McParland, whose efforts to develop the north had earned him the respect and admiration of his colleagues. At the age of forty, he was already very accomplished. Graduating with a degree in mechanical engineering from the University of Toronto, McParland had worked for Noranda Mines before joining Rio Algom Mines as a plant engineer. Rising successively to Chief Engineer, Vice President, and then President and Chief Executive Officer of Brinco, he had prime responsibility for the design and construction of the Churchill Falls Hydro project.[320]

McParland was a great loss to the Institute, having been a founding member and the first chairman of the Mechanical-Electrical Division formed in 1961. When the accident occurred, he was en route to several out-lying Branches across the northeast as a CIM Distinguished Lecturer, a task he had undertaken with considerable enthusiasm. He had declined the monetary award associated with the honour, and was travelling at his

[319] CIM, *Council Minutes*, March 22, 1971, p. 2116.
[320] "Brinco Chief, Businessmen Killed in Crash," *Montreal Gazette*, November 12, 1969. "CIM Necrology," *Bulletin*, no. 693, (January 1970): 17.

own expense.[321] McParland was the sort of individual who could have easily become a CIM president. The tragedy reached deeply into the hearts and minds of his Institute colleagues, and although he could never be replaced, they paid homage by establishing the McParland Memorial Medal in 1972.

In May 1970, another pillar of the Institute passed away. Cam Sproule had suddenly collapsed from a heart attack while giving a speech to the Western Congress of Engineering in Jasper, Alberta, and had died within hours of being taken to the hospital.[322] At the age of sixty-five, he had become one of Canada's foremost authorities on the north, and had been instrumental in bringing together the federal government and private industry to form Panarctic Oils Ltd. Born in Edmonton in 1905, Sproule had studied geology at the University of Alberta and had pursued his doctorate at the University of Toronto. After graduation, he had worked as a consulting geologist before establishing J.C. Sproule and Associates Ltd. in Calgary. A past chairman of the Petroleum and Natural Gas Division in 1953-54, he had also been the first Canadian-born president of the American Association of Petroleum Engineers. As a mark of respect for a man who still had much to contribute, Council established the J.C. Sproule Memorial Plaque in 1973 to honour his "visionary zeal and professional dedication."

The following October, H. Mortimer-Lamb, Secretary Emeritus, passed away at the age of ninety-eight in Vancouver. Besides having served the Institute so ably for many years, he was an outstanding writer, an authority on the mining industry, and a talented painter. After his retirement in 1945, he had devoted much time to his art, receiving public acclaim for his first vernissage. A modest man by nature, he had always considered himself an amateur, and was never able to part with any of his art work. When he died, a piece of CIM's history had gone with him.

A Presidential Succession

James Harrison had followed Paul-Emile Auger, the 1968-69 President. Auger was an economic geologist with degrees from Queen's University, Harvard University and MIT. He had taught at Laval University for almost 20 years before joining Quebec's Department of Mines, and in 1961, he had been appointed Deputy Minister of the Department of Natural Resources. Harrison was also a geologist, with degrees from the University

[321] E.G. Tapp, letter to all Branch Secretaries, November 21, 1969, NAC.
[322] "J.C. Sproule, Panarctic Man, Dies in Jasper," *Montreal Gazette*, May 22, 1970.

of Manitoba and Queen's University, and was known for his keen sense of humour. He had rapidly risen through the ranks of the Geological Survey of Canada and had been appointed its Director in 1956, one of the youngest ever.[323] He later became Senior Assistant Deputy Minister of the Department of Energy, Mines, and Resources. His successor as CIM president was Harold M. Wright. As a young man, Wright had participated as a sprinter in the 1932 Olympic Games. His broad training in geology, geological engineering and metallurgical engineering stood him well when he built Wright Engineers into an internationally recognized consulting firm. In 1993, he was inducted into the Canadian Mining Hall of Fame. Thomas Patching, a professor of mining engineering at the University of Alberta, was next in line as CIM President in 1971-72. A gifted teacher, and authority on coal mining, he had served as chairman of the Coal Division. He would be followed by John H. Schloen in 1972-73. A chemical engineer, Schloen was President of Noranda's Canadian Copper Refiners, and an authority on electrolytic copper refining. With widely-varying backgrounds and interests, these men served CIM during a period of steady and continued progress.

The Foundation

The Canadian Mining and Metallurgical Foundation was inaugurated in 1973 during the Presidency of John H. Schloen, but its origins date back to March 1952, and its legacy is many-faceted. In 1952, Harold M. Wright, then Vice President for District 6, proposed that:

> Consideration be given to the establishment of a fund which might be termed the C.I.M. Foundation. Such a foundation could be developed by a Committee composed of the Past Presidents of the Institute and it could be administered by the Council. . . It could be made an endowment fund for the collection and dissemination of technical and other pertinent information concerned with education for, and advancement of, the mineral industry of Canada.[324]

The proposed CIM Foundation, which had originally evolved out of a discussion between Presidents Oliver B. Hopkins (1951-52) and

[323] Zaslow, 416.
[324] CIM, *Council Minutes*, H.M. Wright, letter to C. Gerow, November 26, 1951, January 4, 1952, p. 335, vol. 7, NAC.

Charles C. Huston (1954-55), and Secretary-Treasurer Gerow (1946-65), was to have eight objectives: to encourage Vice Presidents to visit and address Branches in other districts, to promote Branch welfare by providing speakers and assisting their activities, to encourage Institute membership by mingling with affiliated university groups, to expand library services and provide editorial help, to make abstracts from non-CIM publications available, to arrange for the wider distribution of CIM reprints, and to educate high school students about the mineral industry. These objectives were printed in the *Bulletin*, along with a call for discussion among the CIM membership.[325] The response was dismal: Council received only five letters. Although the general sentiments were in favour of the proposed foundation, there were questions over its objectives.

The following year, a special committee was struck, chaired by Desmond F. Kidd, the 1952-53 CIM President, to investigate ways and means of establishing the foundation, and to suggest a possible scenario for expanded services or special projects. One of the suggestions was to form a trust or foundation fund for tax-deductible donations. Then in 1954, President Arthur E. Flynn (1953-54) suggested that the foundation matter be deferred until after the Sixth Empire Mining and Metallurgical Congress in 1957, as this event would require extensive planning, cooperation from all Branches and technical Divisions, and extra financing. Serious work on the foundation proposal would wait another twenty years.

In 1970, the idea of a foundation gained new momentum, partly due to earlier conversations between John H. Schloen and Gordon Tapp, who had succeeded Gerow as Institute Secretary in 1965. Tapp was familiar with the foundation established by AIME, and hoped to interest Council in building a similar one. Schloen was sympathetic to the idea, and would serve as its first chairman from 1972 to 1991. After being endorsed by the Finance Committee, the proposal was approved by the Executive Committee "to provide the means of creating, developing, and maintaining a program of continuous education for the members of the Institute."[326] The formation of a governing board was recommended to oversee the drafting of appropriate by-laws. Much of this work was taken on by members of the General Committee on Education.

One of the major questions at that time was whether the goals of other existing foundations, such as the Canadian Mineral Industry Educa-

---

[325] "Proposed C.I.M. Foundation," *Bulletin*, no. 479 (March 1952): 128.

[326] CIM, *Executive Committee Minutes*, "C.I.M. Foundation," April 17, 1970, p. 1127.

tion Foundation, overlapped with the proposed CIM foundation. When the General Committee on Education, chaired by René Dufour, concluded that, in fact, the foundations would complement each other, the members of Council unanimously decided to proceed.[327] Dufour, a professor of mining engineering at the Montreal's École Polytechnique since 1961, would become a Charter Member of the Canadian Mining and Metallurgical Foundation (CMMF) in 1973, and its Chairman from 1991 to 1997.

After legal consultation, a draft charter was submitted to the Executive Committee on November 11, 1970.[328] Its eleven aims were both similar and different to Wright's original eight. In addition to providing continuing education for Institute members, there were provisions for student assistance, financial aid to relevant associations, and support for academic institutions. A lecture series would be implemented for Branches, and a mineral industry information program would be prepared for high schools. Funding would rely on donations or bequests, and any excess over current requirements would be invested for future needs. A corporate seal was then designed, bearing the words Canadian Mining and Metallurgical Foundation.

In 1972, President Thomas Patching announced that the Foundation would go ahead as planned. Its purpose was to: "create and support a strong program of educational and charitable activities directed towards improving the contribution of the mineral industry to the progress and well being of the country.[329] In 1973, Council was advised that the CMMF had been established as a legal entity, and an initial request for contributions was in the mail to CIM's Canadian membership.[330]

Unfortunately, the Foundation's financial campaigns were not successful. By 1974, Council understood that it might have to resume financing the Distinguished Lecturers and Visiting Lecturers programs for 1975-76, as well as the activities of the Educational Committee. The CMMF Directors firmly believed that the Foundation should continue to exist in case there was a major benefactor in the future.[331] The event which finally focussed attention back on the CMMF was the National Education Committee Program, an ambitious plan to promote public education about the mineral industry, primarily through elementary and secondary schools. A key element was the Identity Curriculum Project consisting of teaching

[327] CIM, *Council Minutes*, "C.I.M. Foundation, report by René Dufour, Chairman, General Committee on Education," December 14, 1970, p. 2128.

[328] CIM, *Executive Committee Minutes*, November 11, 1970, p. 1175 and 1183.

[329] "C.I.M. Foundation By-laws, Article 1, Purposes," 1972, CIM Archives, Montreal.

[330] CIM, *Council Minutes*, "Canadian Mining and Metallurgical Foundation", April 19, 1973.

[331] CIM, *Council Minutes*, October 27, 1974, minute 22.

materials developed by educators. In 1977, the General Committee on Education appealed to the Foundation for support, and asked it to be responsible for implementing the program. In turn, Council was asked to support the CMMF by providing it with a loan of $50,000,[332] and soon after, it was incorporated as a non-profit organization.

The Program was a success, and by the 1980s, the CMMF had acquired a number of corporate sponsors in addition to donations from the membership. Two of the most important uses of the Foundation's funds continue to be education for the mining industry, and the Distinguished Lecturers Program.[333] The Foundation's title was officially changed in 1993 to the Canadian Mining, Metallurgical and Petroleum Foundation (CMMPF) to reflect CIM's broader constituency and its new name. Two years later, in 1995, the Foundation Endowment Fund was created to give donors a choice of either donating to an endowment, or to the CMMPF general funds. An objective of one million dollars was set for the Fund as a way of commemorating the Institute's 100th Anniversary. Don Worth, CIM's 1996-97 President, is the Foundation's new Chairman.

Tenth Commonwealth Mining and Metallurgical Congress

In 1974, the CIM hosted the Tenth Commonwealth Mining and Metallurgical Congress, which coincided with the Congress's Jubilee Anniversary. As in the past, its main objective was to bring together professionals in the mineral industry throughout the Commonwealth so that they could share their knowledge and experience. The Congress Chairman was John Convey, the recently retired Director of the Mines Branch (1951-1973), who had served as Technical Chairman for the Congress in 1957. Members of the committee included C.C. Huston, Thomas Patching, D.J. Worth and H.M. Wright, all past or future CIM presidents. Convened in Ottawa on September 2nd, the event attracted 837 delegates and their guests from thirty-two countries. The delegates were welcomed by CIM President W.M. Gilchrist, and opening remarks from an array of guest speakers focussed on world energy resources and mineral supplies. A four-

[332] CIM, *Council Minutes*, "A National Education Committee Program," April 21, 1977, p. 2486-2498.

[333] "The Canadian Mining and Metallurgical Foundation - An Asset to CIM," *Bulletin*, no. 969 (April 1993): 7, and *Journal of Canadian Petroleum Technology*, vol. 32, no. 4 (April 1993): 5. "Canadian Mining, Metallurgical and Petroleum Foundation - Your Foundation," *Bulletin*, no. 992, (July 1995): 9.

week excursion followed. With Canadian attendance boosting the delegation above 1,000 in larger centres, travel arrangements eventually required 200 planes, 250 busses, a train, and accommodation in 125 hotels. A farewell dinner was held in Vancouver where the Chairman of the Commonwealth Mining and Metallurgical Institutions, Sir Ronald Prain, pronounced the Congress "a resounding success."[334]

According to Convey, the success of the Congress was "an excellent example of the cooperative force which exists within our parent organization."[335] This was largely due to the generosity of local Branches, some of which exceeded the call of duty. After the opening technical sessions in Ottawa, many delegates took advantage of the special field trips to eastern Canada before flying to Toronto. This meant that some people would have been en route for close to two weeks, and would be weighed down with papers and other material, and longing for a day of respite. In Toronto, members of the Women's Association of the Mining Industry of Canada (WAMIC) and the Toronto Branch met delegates at the airport and assisted them with registration at the Royal York Hotel. One Branch member became an unofficial "postmaster" and helped delegates to lighten their suitcases by wrapping, stamping, and mailing items back to places like Zambia, South Africa, and Australia. Other Branch members invited delegates to bag their laundry and bring it to the hotel reception where they would insure that it would be taken care of by the following evening. The only snag was that when they presented over one hundred bags to the hotel's laundry room, they were informed that the hotel did not provide a laundry service on weekends. Undaunted, the Branch members rose early on Saturday morning and rushed the bags to a "Chinese Laundry" at their own expense, and by 6:00 p.m., all had been returned in good order. The operation had gone so smoothly that most delegates thought it had all been done by the hotel.[336]

Secretariat Moves to Sherbrooke Street

On October 1, 1976, the Institute's headquarters were moved after 61 years in the Drummond Building. The new address was Suite 400, 1130 Sherbrooke Street West, Montreal. "It [was] brand new, high-class and

[334] *Congress in Canada 1974: The Tenth Commonwealth Mining and Metallurgical Congress* (Montreal: CIM, 1974).
[335] CIM, *Council Minutes*, October 27, 1974, minute 21.
[336] Don Worth, interview with author, April 30, 1997.

expensive but the Committees concerned [had] strongly advised better accommodations."[337] CIM's membership was now well over 10,000. Under W. (Mel) Bartley, CIM President in 1976-77, the Institute continued its programs and enlarged on those which provided the greatest benefits. The present concept of district and divisional technical meetings is largely due to Bartley's foresight.

CIM and AIME

From the earliest years, the relationships between AIME and CIM had been amicable and cooperative. In 1959, CIM's Coal Division had established closer ties with the Coal Division of AIME by making its chairman an ex-officio member, and in later years, AIME's Society of Mining Engineering (SME) had often organized joint meetings with CIM. T.E. Caldwell, CIM President in 1979-80, recalls a healthy reciprocal relationship with AIME, particularly at Trail, B.C., where CIM members from Cominco were often welcomed at meetings in Spokane, Washington.[338] When R.R. McNaughton became the first Canadian citizen to be President of AIME in 1961, the relationship continued to flourish. At AIME's Centennial Meeting in New York in 1971, the Institute presented AIME with a special plaque to commemorate the occasion.

A CIM-AIME Liaison Committee had also been in existence for several years, but in 1970, its Chairman reported to CIM's Council that:

> I and other members of the Committee are in agreement that [it] should be dropped. . . there is no point in having a such a Committee since we have done absolutely nothing since it has been established. Should the occasion arise at some future date where firm actions are needed with regard to a problem with AIME, a Committee could be set up at that time to deal with it.[339]

The Committee was dissolved. Concurrently, a news release from the Society of Petroleum Engineers (SPE) of AIME announcing a world-wide membership expansion was brought to the attention of Council. The Petroleum Society of CIM (Petsoc) had an agreement with SPE, which had

337 E.G. Tapp, letter to J.M.Harrison, August 18, 1976.
338 T.E. Caldwell, interview with author, April 30, 1997, Vancouver.
339 CIM, *Council Minutes*, March 22, 1971, minute 48, p. 2121.

The Institute's offices, and library and reading room (right), were in the Drummond Building (above) on St. Catherine Street in Montreal from 1917 to 1976.

Retirement of Dr. R.P.D. Graham (seated), Technical Editor of the Bulletin from 1921 to 1961. From left to right: John I. McGerrigle, Technical Editor and later Editor-in-Chief, CIM Publications, 1962-1983; Professor R.G.K. Morrison, McGill University; John R. Bradfield, CIM President, 1962-63, and President of Noranda Mines Limited; E. Gordon Tapp, CIM Executive Director, 1965-79; G.C. Bateman, CIM President 1940-41; Carlyle (Jerry) Gerow, CIM Executive Director, 1949-63; John Schloen, CIM President, 1972-73; Alfred E. Brewer; and Dr. J.E. Gill, McGill University. In addition to being Technical Editor, Graham had a distinguished career as Professor of Mineralogy at McGill University from 1905 to 1950.

George C. Mackenzie, 1920-31

E.J. (Curly) Carlyle, 1931-49

Carlyle (Jerry) Gerow, 1949-63

Richard E. (Dick) Barrett, 1963-65

CIM Secretary Treasurers and Executive Directors

E. Gordon Tapp, 1965-79

Gordon F. Skilling, 1979-82

Pierre Michaud, 1982-92

Yvan Jacques, 1992-

CIM Executive Directors

ANNUAL
DINNER
OF THE
CANADIAN
MINING INSTITUTE
THE CANADIAN MINING INSTITUTE
MONTREAL
FRIDAY MARCH 5th 1909

"A bunch of the boys whooping it up" at the Smoker at the 30th Annual General Meeting of the Institute held at the Royal Alexandria Hotel in Winnipeg in 1929.

The Annual Dinner at the 30th CIM Annual General Meeting. Black-tie dress was de rigueur.

The Annual Ball receiving line at the 74th AGM in Ottawa in 1972. Among those in the photograph are AGM Chairman D.G. MacKay (second from left), CIM Secretary-Treasurer Gordon Skilling (sixth from left), and Alexis Ignatieff (tenth from left).

Passing the torch. Outgoing President Peter Tarassoff of Noranda Inc., and incoming President Bill Stanley of Coopers and Lybrand at the Institute's Annual Business Luncheon at the 93rd Annual General Meeting in Vancouver in 1991.

Nine Past Presidents at the 84th Annual General Meeting of the Institute held in 1982 in Quebec City. From left: R.D. Lord (1977-78); P.D. O'Connell (1975-76); A.O. Dufresne (1950-51); H.M. Wright (1970-71); D.E.G. Schmitt (1980-81); J.H. Schloen (1972-73); W.M. Gilchrist (1974-75); E.H. Caldwell (1979-80); and J.P. Drolet (1978-79).

Past Presidents at the 99th Annual General Meeting in Vancouver in 1997. Seated, from left: J.J. Laffin (1989-90); W.J. Riva (1984-85); G. Massobrio (1995-96); and W.M. Gilchrist (1974-75). Standing, from left: E.H. Caldwell (1979-80); R. Dufour (1988-89); D.S. Robertson (1993-94); H.M. Wright (1970-71); P.V. Young (1983-84); W.E. Stanley (1991-92); and W. Almdal (1994-95).

been negotiated in 1966 by CIM President George W. Govier, under which each society was prohibited from campaigning for members in the other's country. The news release was clearly at odds with this agreement. Council agreed to observe the situation for a while before taking specific action.[340]

In 1973, the matter became more serious. SPE reported that it had cancelled its reciprocal agreement with CIM's Petroleum Society. CIM Council noted that "the parent AIME organization was not in sympathy with the ambitions of the SPE but that it was probably powerless to bring an end to the announced objectives of SPE."[341] Both the Petroleum Society and SPE were striving to increase membership, and Petsoc, led by Wayne Glenn, an American who was also a prominent member of SPE, did its best to swing interest in its favour.[342]

The situation grew more tense in the years ahead. In 1989, the Petroleum Society's Chairman-elect, Stephen K. Gordon, raised its concerns with CIM President Peter Tarassoff about the establishment of an SPE Section in Calgary. Because the "SPE provides services which closely parallel those of the Petroleum Society," it was becoming increasingly difficult to remain neutral.[343] The Petroleum Society realized that it needed to adopt a new strategy to compete with the SPE. It had already established its own office in Calgary, managed since 1986 by Catherine Buchanan, editor of the *Journal of Canadian Petroleum Technology (JCPT)*. The next step was to transfer all the *JCPT* operations from Montreal to Calgary in 1993.[344] This provided the Petroleum Society with a strong presence in Calgary, and sent a clear message to prospective Canadian members that it was the organization of choice.

Since Govier's term as CIM President in 1966-67, five other members of the Petroleum Society have served as Institute presidents: Patrick O'Connell (1975-76), Noel Cleland (1981-82), Lindsay Milne (1987-88), A.D. (Tony) Anton (1992-93), and William Almdal (1993-94). Each would act as a link between the Petroleum Society and CIM, and contribute greatly to the Institute's progress and stability. Over the years, the Petroleum Society has become one of the Institute's largest technical divisions, and its

---

340 Ibid., minute 55, p. 2122.

341 CIM, *Council Minutes*, April 19, 1973, minute 9, p. 2372.

342 George W. Govier, interview with author, November 29, 1996, Calgary.

343 S.K. Gordon, Chairman-Elect, Petroleum Society, letter to P. Tarassoff, CIM President-Elect, September 22, 1989.

344 "Alberta - A History of Growth: the Petroleum Society of CIM - A Short History," *CIM Directory* (1981): 85-96. Barbara Howes, "Everything you Wanted to Know About the Petroleum Society But Were Afraid to Ask!" *JCPT*, vol. 33, no. 7 (September 1994): 11-12. S. Aubrey Kerr, "CIM and Oil," unpublished manuscript, October 24, 1995, CIM Centennial files. Catherine Buchanan, interview with author, November 15, 1996.

annual technical meetings have gained a strong following, and contributed to the Society's financial strength. The Society would later help the Institute to overcome some of its own financial difficulties, but some serious questions of autonomy would remain.

A New Executive Director and New Challenges

In 1979, Gordon Tapp retired as CIM's Executive Director. At his last official Annual General Meeting in Montreal, he was awarded a special scroll by the Institute's Past Presidents, in honour of his exemplary service."[345] His contributions were also recognized by other CIM constituencies, and the Metallurgical Society presented him with a painting. Tapp's successor was Gordon F. Skilling, who had been Secretary-Treasurer for the past fourteen years, and was now appointed Executive Director.

Skilling was a geologist who had worked in hard rock mining before accepting an editorial position in 1955 with *Canadian Oil and Gas Industries*, a publication of National Business Publications Limited.[346] As CIM Secretary-Treasurer, he had been assisting Gordon Tapp and John I. McGerricle, Technical Editor, with the *JCPT*. Skilling's major strength was publishing, and he knew the business thoroughly. Along with increasing sales, subscriptions, and advertising in the *Bulletin*, he had introduced the very successful *CIM Directory* in 1967, and the *CIM Reporter* in 1975.

The Institute's growing need for change would place new demands on Skilling, whose experience had led him to follow in Tapp's footsteps. From 1980 onwards, many of the issues and projects undertaken by the Institute would require a team approach to problem-solving, and several years of study and dialogue to implement. The Institute was expanding rapidly, and fresh ideas were pressing the Council to take new risks and to challenge some of the older traditions. In 1981, the Canadian Mineral Processors (CMP) had joined the Institute, bringing the total number of technical Divisions and Societies to eight. Communications between the Council, the Secretariat, and the Institute's many and diverse constituent groups were becoming more demanding. With 57 Branches in existence, it was next to impossible for the President and Executive Secretary to visit more than a small fraction in a year, and the Institute's total membership

[345] E.G. Tapp, interview with author, July 7, 1996, Belleville, Ontario.

[346] S. Aubrey Kerr, "COGI/OTS," draft no. 4, November 25, 1994, 8 p. The publications of the *Canadian Oil and Gas Industries* (COGI, 1949-63) and *Oilfields Technical Society* (OTS) were owned by National Business Publications Limited and became staunch competitors of CIM's *JCPT*.

had exceeded 12,000. Even the Annual General Meeting had grown beyond its means, and organizing a meeting for over 3,000 delegates was taxing for the Program Committee, the Secretariat, and the host Branch. This was sometimes leading to friction over budgets and financing.

For Gordon Skilling, the stress of coping with the all the demands was taking a toll, and in March 1982, he asked Council to be relieved of the Executive Directorship and to continue in a less demanding position.[347] Unfortunately, his health would not allow him the opportunity, and he retired soon afterwards. His successor was Pierre Michaud, a mining engineer, and the first French Canadian to hold the position.[348]

Pierre Michaud, Executive Director

Like his predecessors, Pierre Michaud was already well-known to the Institute. A CIM member since 1965, he had served as Chairman of the Montreal Branch in 1979-80, and as a Councillor for District 2 in 1981-82. He had spent seven years in supervisory positions with East Sullivan Mines Limited, a copper-producer in Val d'Or, Quebec, before joining Atlas Copco in 1957, where he became marketing manager in 1981. Michaud would be Executive Director for the next ten years, a turbulent time for both the Institute and the mining industry. On the eve of his retirement, he described the organization that he had grown to know extremely well:

> The Institute is a strong association, stronger than most people believe. It draws its strength from the hundreds of volunteers who succeed themselves like an endless chain, forever working their hearts out to fulfill the Institute's vocation which fundamentally is to improve the members' status within the mining, metallurgical and petroleum corps of professional workers.[349]

It was with considerable sadness that CIM colleagues learned of Pierre Michaud's death four years later, on April 26, 1992. Of the many tributes from Institute Past Presidents printed in the *Bulletin*, this passage was particularly fitting:

[347] CIM, *Council Minutes*, "Executive Director Replacement," March 19, 1982, minute 41.
[348] "Pierre Michaud: Personality Profile," *Bulletin*, no. 838 (February 1982): 36-40. *Council Minutes*, October 6, 1982, "Ratification of New Executive Director, CIM," minute 10.
[349] Pierre Michaud, "The Final Word," *Bulletin*, no. 963 (September 1992): 9.

> Pierre distinguished himself in his role as Executive Director by not only representing the Institute in a very professional and competent manner, but beyond that he brought a sensitivity to his dealings with the various factions of CIM that was very apropos at the time. . . he will long be remembered for his strong leadership of the CIM organization and his uncommon touch in dealing with a diverse organization.[350]

Michaud had carried out his duties with pride and respect.

The Financial Crisis

On the surface, the Institute appeared to be in a strong financial position when it recorded a $311,718 excess in revenue over operating expenses in 1980.[351] The *Bulletin*, *Directory*, *Reporter*, and *JCPT* were in good shape, except for a worrisome loss in advertising revenue, which was expected to rebound the following year. However, the Institute's reserve account had suffered heavy losses from 1974 to 1978 as a result of a devaluation of the Canadian dollar, and additional funds had been needed to build-up the reserve account in both 1979 and 1980. The Finance Committee had also pointed out that:

> if inflation continues at the same rate as the last few years the budget for the Institute will reach 3 million by 1985, and the 390,000 working capital in the current budget and projected through to 1985 would not be adequate. .[352]

Council was impressed by the five-year forecast provided by the Finance Committee, and agreed that it would be a valuable tool for assessing future needs. As in the past, the Institute relied heavily on advertising revenue from its publications, and as long as membership continued to increase, each would provide a steady income. What Council could not predict was the loss of an important publication staff member, a change in Executive Directors, and the depth and duration of the recession.

---

[350] Lindsay Milne, in: René Dufour, "In Memoriam: Pierre Michaud," *Bulletin*, no. 1003 (September 1996): 8-9.

[351] CIM, *Council Minutes*, "Financial Statements to August 31, 1980," September 17, 1980, minute 19.

[352] Ibid.

On January 18, 1983, the Institute's Editor-in-Chief, John McGerrigle, died unexpectedly. He had been with CIM since 1962, and as Assistant Technical Editor, he had worked closely with R.P.D. Graham until the latter's retirement. McGerrigle continued to increase his knowledge about publishing, assuming the position of Editor-in-Chief in 1979.[353] After Gordon Skilling resigned in 1982, McGerrigle emerged as CIM's publications expert. His premature death meant that the staff had to quickly step in, learn the procedures, and guide the work-in-progress towards the press deadlines. Much of the added responsibility was assumed by Pierre Michaud and Perla Gantz, an editorial assistant who would be promoted to Assistant Editor in 1986, and then to Editor in 1990.

McGerrigle's death was just the start of a long and difficult year. The mining industry was beginning to feel the effects of a recession, and the resulting large drop in advertising revenue forced Council to reduce the Secretariat permanent staff from 34 to 26 members. For the first time in recent memory, the Institute recorded an operating loss and had to borrow from its reserve funds. Publications were scaled back, and as the recession deepened, only 1,613 people attended the 1983 AGM in Winnipeg, compared with 2,534 in Quebec City the previous year, and 3,553 in Calgary in 1981. Six Branches had also ceased operations because of economic pressures and mine closures.

Council debated raising membership dues, and on January 1, 1984, they were increased by 20% for all categories except for students.[354] Then in 1987, Council acted on a recommendation from the Finance Committee, and another dues increase was approved for January 1, 1988. When yet another raise in dues was suggested, this time to be effective January 1, 1991, the Council under President René Dufour decided to refer the matter to the incoming Council for study.[355] Although the Institute's finances were still in the black, the proposed dues increase was clearly a 'red flag' pointing to a more serious underlying situation. By early 1990, the Finance Committee chaired by Réal Lafleur was projecting an operating deficit for the year, and forecasting that the deficit would grow in 1991 and 1992. Alarmed, the 1990-91 Council directed that a Task Force on CIM Finances be established to review CIM's revenues and expenditures, identify ways of reducing costs

[353] "John McGerrigle, In Memoriam, 1933-1983," *Bulletin*, no. 850 (February 1983): 7. His father was H.W. McGerrigle, an outstanding geologist and CIM member in whose honour the McGerrigle Mountains of Gaspé were named.

[354] CIM, *Council Minutes*, "Institute Dues Increase," September 19, 1983, minute 21.

[355] CIM, *Council Minutes*, "Dues Increase," March 20, 1989, minute 6.

and increasing revenues, and study alternatives to the existing dues structure.[356] C. Bruce Ross, CIM's 1986-87 President, agreed to chair the Task Force. In its report on February 11, 1991, the Task Force recommended tighter control of AGM finances, higher registration fees for all CIM meetings, and an annual dues increase to reflect the rate of inflation.[357] With a continued deterioration of the Institute's finances, Council then directed the President, Peter Tarassoff, to appoint a new committee to review the projected revenues and expenditures for 1991, recommend ways of achieving a balanced budget in 1992, and develop an action plan for implementing the recommendations of the Ross task force.[358] This committee was chaired by Len Surges, the Chairman of the Canadian Mineral Processors. In what evolved into a lengthy and detailed analysis of CIM's operating structures, with many recommendations, the message of the Special Committee on CIM Finances was clear: that "Council take strong and decisive action now to restore the health of the Institute."[359]

In the early years of the Institute, the Branches and technical Divisions and Societies had been assisted financially by the Institute when it was needed. Now it was their turn to help the Institute to reduce its operating deficit. One of the first to offer support was the Montreal Branch, which agreed to a reduced share, from 50% to 25%, of any surplus revenue accruing from the 1992 AGM, to be held in Montreal.[360] Many other Branches, Societies, and Divisions also responded with donations from their coffers. By January 1992, CIM President W.E. (Bill) Stanley could announce that the response to the Institute's appeal for funds had been "heartwarming", and that the donations totalling $177,000 would help "to see CIM through its difficult timers."[361] The Institute in turn had taken steps to reduce labour costs, improve the return on its investments, and increase membership. A dues increase for 1992 was also brought in.[362] But even with all these measures, the deficit had again increased. In an effort to further reduce costs, the *Bulletin* was shortened by 92 pages, and the December 1991 issue was cancelled. Subsequently, the *Bulletin* was reduced to ten issues a year.[363]

---

[356] CIM, *Council Minutes*, June 11, 1990, minute 15. P. Tarassoff, letter to C.B. Ross, June 20, 1990. "Highlights of the Calgary Meeting," *Bulletin*, no. 946 (February 1991): 7.

[357] CIM, *Council Minutes*, "C.B. Ross, letter to P. Tarassoff, January 25, 1991," February 11, 1991, minute 5.2.

[358] P. Tarassoff, letter to L.J. Surges, February 14, 1991.

[359] *Report of the Special Committee on CIM Finances*, draft, July 10, 1991. CIM, *Council Minutes*, "Report from Task Forces on Institute Finances," September 30, 1991, minute 8.

[360] CIM, *Council Minutes*, "94th Annual General Meeting," September 30, 1991, minute 10.3.

[361] *Communiqué*, vol. 1, no.8 (January 1992).

[362] Notice to all CIM Members - Annual Dues Increase," *Bulletin*, no. 952 (1991): 7.

[363] W. E. Stanley, "Report of Council – 1991, President's Report," *CIM Directory* (1992): 43.

Although great efforts were made to overcome the Institute's financial difficulties, the deficit would persist for six more years. In fact, the worst was yet to come. By the end of 1992, advertising revenues continued to spiral downwards, and even with a membership increase of 3%, there was another shortfall. In fact, the Institute was a microcosm for what was happening throughout Canada in the 1990s. In his President's Report, Tony Anton explained that:

> Our industry continued its sluggish performance for the best part of 1992, with very little recovery from the recessionary times. . . Resource companies, in their attempt to improve their economic performance, continued to downsize resulting in more people losing their jobs.[364]

The following year, the deficit had more than doubled, but Council remained optimistic for a turnaround. In his *Bulletin* column, "Observations of the President," David S. Robertson (1993-94) said:

> I am confident that despite the pressures experienced by our industry in environmental regulation, base metal prices, and declining reserves, there will be bettter days ahead. . . At CIM as elsewhere, we are being forced to make those sacrifices (for better days in order) to survive.[365]

By 1994, the operating deficit was a staggering $300,023, the highest ever recorded in the Institute's history. However, there were some positive signs that the downward trend was reaching an end. The largest ever AGM was held in Toronto in 1994, in conjunction with the largest ever Trade Show, membership had risen, and advertising revenues were back on-track.[366] By 1995, a slight excess was reported in operating expenditures over operating revenues, but an overall shortfall remained. During the year, the Secretariat had undergone restructuring, and membership dues were raised.[367] Finally, in 1996, a surplus of $122,684 brought the Institute's finances back into line. Although the struggle had been long and arduous, the crisis had been overcome. A major change was also made in how the

[364] A.D. Anton, "Report of Council - 1992, President's Report," *CIM Directory* (1993): 31.
[365] D.S. Robertson, "Observations by the President," *Bulletin*, no. 973 (September 1993): 7.
[366] W. Almdal, "Report of Council - 1994, President's Report," *CIM Directory* (1995): 35.
[367] "Notice to all CIM members: Increase of Annual Fees," *Bulletin*, no. 978 (March 1994): 10.

Institute would monitor its financial situation in the future. The Finance Committee was disbanded but the Chairman remained on Council, and Council itself assumed responsibility for reviewing financial statements. In his 1996 Presidential Report, Don Worth commented: "C'est toujours plus interéssant leading an organization when it is in the black."[368]

Breaking With Tradition

During the restructuring process, there was another change at the Secretariat. When Pierre Michaud retired on September 1, 1992, he was replaced by Yvan Jacques. Jacques was the first Executive Director to be hired from outside the mining industry. He had acquired a background in science, and an MBA from the University of Sherbrooke. Having worked as an Assistant Deputy Minister for the Quebec Ministry of Natural Resources, he had considerable knowledge of the working of governments, and as Executive Vice President for the Agricultural Institute of Canada from 1988 to 1992, he was familiar with professional institutions.[369] He brought new energy and expertise to the position of Executive Director.

Improving the Bulletin

By 1994, the *Bulletin* Committee, chaired by John Udd, was ready to report on the dilemma of how to maintain the *Bulletin* as a world-class journal, yet adapt to the new realities of corporate advertising. The committee arrived at three basic recommendations. The first was to discontinue the printing of abstracts of papers that were to be presented at meetings and, instead, publish only their titles and the authors' names. Because the abstracts were often reproduced in conference programs, this represented a substantial cost saving. The second was to issue pages quotas to the Societies and Divisions for technical papers and activity reports, and for Institute committee reports, Branch and Institute news.

Page quotas were certainly not new, having been hotly debated by the technical Societies and Divisions in the early 1960s, when they had often been overruled. It was now time to account for every printed page. This would also balance the *Bulletin* in terms of technical and non-technical content. The third recommendation concerned deadlines; late sub-

[368] D.J. Worth, "Report of Council - 1996, President's Report," *CIM Directory* (1997): 74.
[369] Yvan Jacques, "New Executive Director of CIM," *Bulletin*, no. 964 (October 1992): 10.

missions would no longer be accepted. While the decisions on abstracts and deadlines were firm, the Committee emphasized that the quotas could be amended as necessary. These measures helped the Institute to make some gains during its financial crisis, but their success also meant that they were permanent.[370]

Modernizing the Secretariat at the Xerox Tower

Although computers had been commonplace in the business world for some time, and there had been a CIM Process Control and Computer Applications Committee since 1973, the Institute took a piece-meal approach to modernizing its own Secretariat. It was not until the 1990s that the Secretariat would be brought into the information age. In the previous decade, the Institute's clerical work had increased dramatically. Membership in 1980 stood at 11,818, the highest ever recorded in the Institute's 83-year history, but this level would prove difficult to maintain in the face of later recessions, and downturns and restructuring in the industry. When CIM's 85th anniversary was celebrated at the 1983 Annual General Meeting, which was held in Winnipeg for the first time, 1,613 delegates were in attendance.

That same year, the Canadian National Committee on Rock Mechanics, formed in 1974, had applied for division status, but Council decided it had too few members.[371] It was then renamed the Rock Mechanics and Strata Control Committee. Discussions had also been ongoing, since 1973, about giving the Mineral Economics Committee divisional status; this was to be realized only in 1992, with the formation of the Mineral Economics Society. In 1985, the name of the Mechanical-Electrical Division was changed to the Maintenance/Engineering Division. These changes, while necessary and commendable, placed additional demands on the Secretariat.

The year 1989 saw another move of the Secretariat offices. After thirteen years at 1130 Sherbrooke Street West, new offices were secured in the Xerox Tower in the Alexis Nihon Plaza at the corner of Atwater and Ste. Catherine Streets. While some modernization had been introduced in the Secretariat's publishing department, the daily functioning of the office was still mostly done manually. When John Laffin reviewed the situation as CIM President for 1989-90, he was astounded that "there was still no word processing, and that people would have to go to the filing

[370] John Udd, Chairman, "Report from the Bulletin Committee," *Bulletin*, no. 976 (January 1994): 7.
[371] CIM, *Council Minutes*, "Request by the Canadian National Committee on Rock Mechanics for Change of Status from Committee to Division," October 6, 1982, minute 30.

cabinets to retrieve information" on CIM members.[372] Laffin, who was then chairing a Secretariat Review Task Force, had retained an independent firm to assess the Secretariat's needs for an information system.[373] After consultation with the Executive Committee, a decision was made to install a Local Area Network (LAN) computer system at a cost of approximately $65,000. This system became operational by mid-December, 1989, but purchasing new computers and investing in software was only half the problem. The office staff would also need to be retrained, and new procedures adopted. Long-time CIM employees devoted many extra hours to mastering the necessary skills.

Although there were signs of the Institute's imminent financial troubles, and expenses were beginning to be curtailed, Council realized that stronger efforts were needed to increase the membership. A manager was hired to concentrate on membership recruitment and development, and to arrange for a program of continuing educational courses and workshops. Council also accepted an offer from the Metallurgical Society (MetSoc) to cover the cost of hiring an Administrative Assistant to deal exclusively with MetSoc's administrative needs.

As technology changes, there is a constant need for adaptation if an organization is to maintain efficiency and effectiveness. But it always takes effort and confidence to embark on new ways of doing things. At the Secretariat, desktop publishing was introduced in 1995.[374] A new computer system was installed in 1996, and a communication system featuring electronic as well as voice mail was put into operation. Office procedures were modernized, and quality management was implemented.[375] CIM took yet another step forward when in 1996 it established an interactive web site on the World Wide Web as an-industry sponsored Centennial project. A vast amount of information on CIM's multi-faceted activities was now available online. The Institute's first secretary would surely have been astonished to know that all the information contained in the Institute's library in 1898 could now be stored in a few CD-ROMs that would easily fit in his coat pocket.

The remake of the Institute's Secretariat was completed in 1997 when it was renamed the National Office.

[372] John Laffin, interview with author, April 29, 1997.
[373] Report to Council, "Progress Report, Task Force on Secretariat Workings," October 25, 1989.
[374] W. Almdal, "Observations by the President," *Bulletin*, no. 990 (May 1995): 6.
[375] Giorgio Massobrio, "Observations by the President," *Bulletin*, no. 999 (April 1996): 8.

Four Classes of Membership

Since the beginning of the Institute, by-law amendments had been frequent occurrences. Of the many changes in the modern era, two of the most significant were the reclassification of members, and the reduction in Council. Although there had been a By-law Committee since 1981, it was not very active until 1987, when Council reaffirmed its existence and asked that the By-laws be reviewed.[376] This process had been initiated by a request spearheaded by the Maintenance/Engineering Division. Steven Harapiak, CIM's President in 1985-86, volunteered to chair the Committee, which consisted of representatives from each of the eight Divisions and Societies, and the Institute's Executive Director. Council urged the Committee to present its recommendations within a year.

C. Bruce Ross succeeded Harapiak as Chairman in May 1988.[377] After reviewing information compiled since 1981, the Committee presented a preliminary report to Council in March 1989. Michaud, who was the Committee's secretary, expressed the opinion that "the By-laws were too important to the Institute to be rushed through for approval and that they would probably be approved by Council at [its] March 1990 [meeting]."[378] The exception was the suggestion by the Membership Committee that Council pre-approve the proposed new membership categories in time for CIM's recruitment brochure. The Associate Member and Associate Junior Member classes were dropped, leaving four basic categories: Member, Junior Member, Corporate Member, and Student Member. In August 1990, a ballot was circulated asking the membership to approve a series of by-law amendments, including ones which dealt with the revised member definitions. These were accepted and the new CIM By-laws were published on September 30, 1990.

Earlier, in May 1986, the Metallurgical Society had requested Council to consider a Fellowship category for CIM as a mark of distinction. The proposal was that Fellows would be elected according to specific guidelines, and their number would be limited to a small proportion of the membership.[379] The final proposal was developed by the Chairmen of the

[376] CIM, *Council Minutes*, "By-Law Revision Work," March 29, 1981, minute 41, and "Approval of By-Law Revision Committee," December 10, 1987, minute 8.

[377] CIM, *Council Minutes*, "By-Laws Revision Committee," May 1, 1988, minute 7.

[378] CIM, *Council Minutes*, "Report from the By-Law Revision Committee," March 20, 1989, minute 8.

[379] CIM, *Council Minutes*, "The Metallurgical Society of CIM - Fellowship Category," March 17, 1986, minute 39. This was not the same as the term "Fellow," introduced by Mortimer-Lamb in 1922, which would have included all the Institute's professional members.

technical Divisions, Societies, and Committees, and presented to Council in the fall, and adopted in 1987.

Reorganization of Council

The problems of having a cohesive and representative Council can be traced throughout CIM's history. In 1940, E.J. Carlyle, then Secretary-Treasurer, stressed at some length:

> the desirability of securing more representative attendance at Council meetings. During a recent period of four consecutive Council-years, 107 members were listed as Councillors, and of that, 64 did not, and doubtless could not attend a single meeting of Council during their terms of office. Occasionally an entire year would pass with one or two provinces never represented. This made adoption of national policies difficult.[380]

Gradually, over the next thirty years, changes were made as the new technical Divisions and Societies came into focus. In 1963, Council seats for the technical groups would increase to two members each, and those for Districts would be reduced from five to four. By 1970, Council totalled forty-eight members. Four years later, during the presidency of Patrick D. O'Connell (1974-75), CIM organizational charts and questionnaires were distributed to all former District Vice Presidents, who subsequently formed an Ad Hoc Committee. Its goal was to improve communications within the Institute, and increase representation on Council for members. One immediate outcome was the addition of District Vice Presidents to the Executive Committee until the end of the 1976 term.[381] In its final report, tabled and read to Council by O'Connell on May 8, 1975, the Committee expressed the view that:

> the time may come when it will make sense to eliminate the geographical structure of the Institute; in other words eliminate the need for District and Vice President and Councillors. . .[it] should forget about geographical repre-

[380] CIM, *Council Minutes*, April 13, 1940, p. 377, vol. 4, NAC.

[381] CIM, *Council Minutes*, "Institute Organization," October 27, 1974, minute 23, and "Institute Organization," March 14, 1975, minute 40.

sentation and growth and concentrate on membership increases via its Division/Societies.[382]

The President-elect, W. (Mel) Bartley, on assuming responsibility for the Ad Hoc Committee, pointed out the need for By-law changes, and "promised to go to every grass root for points of view before reporting to the next meeting of Council."[383]

Although the recommendations to eliminate Districts did not materialize, calls were made for the position of District Councillor to be more meaningful and functional. The lengthy wrangling over the addition and removal of councillors of every description continued into 1976, after which the business of Council reorganization would be put on hold for another sixteen years.

In the interim, in 1977, Council decided to combine the executives of District 6 and the historic British Columbia Section under Districts 6.[384] By this time, Districts 4 and 6 were organizing their own technical meetings and awards.[385] The legacy of the B.C. Section dated back to the original British Columbia Branch, formed in Nelson, B.C. in 1902. It was later renamed the Western Branch (1908), the British Columbia Division (1921), and then the B.C. Section (1945), and once enjoyed a similar position on Council to that of the MSNS. When the district structure came into effect in 1922, District 6 assumed responsibility for British Columbia, although most of the arrangements for the Annual Western Meeting and other activities continued to be handled by the B.C. Section. This meant that the role of the District Vice President was often diminished. The overlap was finally eliminated when the B.C. Section amalgamated with the Vancouver Branch in 1980. According to E.H. (Ted) Caldwell, who had been the Chairman of the B.C. Section in 1963-64, "it was a tremendous change and improvement as it brought the geographic areas together. . . there was no real opposition, and the transition went fairly easily.[386] W. (Mel) Bartley, CIM President at the time, firmly believed in the benefits of District technical meetings; the Mel W. Bartley Award, later named in his honour, recognizes Branches which best exemplify the aims and objectives of the Institute.

---

[382] CIM, *Council Minutes*, "Institute Organization," May 8, 1975, minute 8.
[383] Ibid.
[384] CIM, *Council Minutes*, "Combine District No.6 and B.C. Section," April 21, 1977, minute 67.
[385] "CIM Report of Council - 1977," *CIM Directory* (1978): 55.
[386] E.H. Caldwell, interview with author, April 30, 1997, Vancouver.

In 1992, Council requested President A.D. (Tony) Anton to strike a committee representing all components of the Institute. Its mandate was to develop recommendations on altering the composition of Council.[387] Under the Chairmanship of Steve Gordon, the committee recommended that Council be reduced from 53 to 20 voting members. Council would now consist of the Presidential "triumvirate," six District Vice Presidents, the Chairmen of the technical Societies and Divisions, the President of the MSNS, and the Chairman of the Finance Committee. The District Council structure would remain intact, and in each of the six Districts, it would consist of the District Vice President, four District Councillors, and Branch and Section Chairpersons.[388] The overall result was a better focussed Council, and more satisfactory representation at the District level.

The Environmental Challenge

In his Presidential address at the 74th AGM in Ottawa, 1972, T.H. Patching introduced an environmental dilemma:

> During all of the discussion of environment and pollution standards in recent years there has not been as much attention to the possibility that instead of our way of life being jeopardized by pollution, it may some day be jeopardized by shortages of necessary materials, and more especially, by a lack of suitable [energy] sources to sustain it.[389]

Even though there were no visible signs of mineral or energy shortages in the early 1970s, Patching, like others, believed that inevitably, Canada would deplete its supplies of non-renewable resources. Because mineral and fossil fuel deposits were non-renewable, and exploration for new reserves was slowing down, it made perfect sense to create more efficient processes for extraction and utilization. Patching concluded by stating, "It has not been my intention to try to create a scare about future supplies of materials and energy. . . but we do have to look ahead and plan wisely."[390]

---

[387] CIM, *Council Minutes*, "Institute Reorganization and Council Reform," October 19, 1992, minute 10.6.

[388] A.D. (Tony) Anton, "Observations by the President," and "Existing By-Laws and Proposed Amendments," *Bulletin*, no. 968 (March 1993): 8-11.

[389] T.H. Patching, "The Accomplishments of CIM and the Challenges to Industry," *CIM Directory* (1972): 46-48.

[390] Ibid.

Canadian interest in the environment has basically two roots: the old conservation movement which emerged in the 1890s, and the new conservation movement which resurfaced in the 1960s. The former was closely bound to the concept of "wise use," as practiced in 1906 by the Canadian Forestry Association.[391] When Canadian delegates attended the North American Conservation Conference in 1909, they accepted the American call for the establishment of a Canadian Commission of Conservation (1909-1921). Under Clifford Sifton, the federal government's dynamic Minister of the Interior, the Commission emerged as a research-based organization, which included university people, like Frank Dawson Adams, among its membership. Its varied mandate included minerals and fuels.[392] But just as the Commission was beginning to exert its influence, charges were made that it was duplicating other government services, and it was cancelled in 1921. For the next forty years, little happened within government to monitor industrial development and agricultural expansion.

Between 1967 and 1970, the dramatic surge in public concern over pollution, sparked by dire journalistic predictions, ushered in a new environmental age.[393] During this period, Pollution Probe (1969) was formed at the University of Toronto, environmental studies began to appear in colleges and universities,[394] and Environment Canada was created by the Federal Government (1971). Although the "grass roots" movement remained intact, support from the general public dwindled over the next ten years due to widespread fears that pollution control measures would curtail economic growth and lead to high unemployment.[395] Consequently, Environment Canada had little decision-making influence until 1986.

The major turning point was the Bruntland Report of the United Nations World Commission on Environment and Development, which stimulated a shift in global politics, economy, and trade.[396] What followed in Canada was the National Task Force on the Environment and Economy, which stated that "environment and economic concerns" must coincide.[397]

---

[391] Thomas L. Burton, *Natural Resource Policy in Canada. Issues and Perspectives* (Toronto: McClelland and Stewart, 1972).

[392] Ibid., 33.

[393] Paul M. Brown, "Organizational Design as Policy Instrument," in *Canadian Environmental Policy: Ecosystems, Politics, and Process*. Robert Boardman ed.(Oxford, U.K.: Oxford University Press, 1992).

[394] E. Tina Crossfield and Peter Barry, "The Growth of Environmental Studies Programs at Universities," conference paper, Canadian Science and Technology Historical Association, Kingston, October 1992.

[395] "Pollution Problem Serious," *Toronto Star*, June 19, 1989, p. A1, A11.

[396] United Nations World Commission on Environment and Development. Bruntland. May 1986.

[397] Boardman, 33.

As government and industry began to agree on strategies for sustainable development, it became easier for the public to support their efforts.

Canada's mining and metallurgical industry was certainly not a stranger to environmental problems. As early as 1867, farmers in Quebec's Eastern Townships had lodged complaints against the practice of roasting copper ore in open heaps "which destroys not only the forests and farm crops and injures our cattle but its dense and poisonous gases enter our houses."[398] In 1879, the Orford Nickel and Copper Company had "to pay large sums of money yearly to neighbouring farmers for damage done to crops and grasses by the sulphurous gases produced at the works."[399] Smelter fumes were also a problem at Sudbury, where, in 1925, the first sulphuric acid plant was built at the Mond Nickel Company's Coniston smelter.[400]

Around the same time, Cominco's smelter at Trail, B.C., was faced with a major air pollution problem as damage claims were filed by farmers in neighbouring Washington State. The problem was tackled by future CIM President Randolph W. Diamond, and in 1931, a large-scale chemical fertilizer plant was placed in operation.[401] This pattern of converting a sulphur-dioxide pollutant into a saleable product was to be repeated elsewhere in the industry. By the 1950s and 1960s, new metallurgical technologies were emerging which were more productive, more energy efficient, and more friendly to the environment.

Since the early 1990s, many voluntary environmental initiatives have taken place within the Canadian mining industry.[402] Two notable programs are the Mine Environment Neutral Drainage (MEND) Program, and the Accelerated Reduction/Elimination of Toxins (ARET). Formed in 1989, MEND is responsible for research leading to prediction, prevention, control, and long-term management of acid generating tailings and waste rock. Composed of representatives from the mining industry, and various government departments, many potential environmental problems have been eliminated. ARET, introduced in 1994, includes 92% of Canadian companies involved in base metal production, and all non-ferrous smelters, along with health and academic associations, and federal and provincial govern-

398 W. Gillies Ross, *Three Eastern Townships Mining Villages in Quebec, 1863-1972: Albert Mines, Capelton*, Eustis (Sherbrooke: GGC, 1996) p. 58.
399 Ibid., 59-60.
400 C.J.S. Warrington and R.V.V. Nicholls, *A History of Chemistry in Canada* (Toronto: Pitman and Sons, 1949) pp. 49-51.
401 Ibid., 51-53.
402 Andrew Oliver, "Mining, the Environment and CIM," *Bulletin*, no. 989 (April 1995): 19-20. Bob Irvine and Elizabeth Gardiner, "CIM Environmental Policy," *Bulletin*, no. 1010 (May 1997): 8-9.

ment departments. The program projects a 70% reduction in the emission of twelve harmful substances by the year 2000.

CIM and the Environment

The CIM Environment Committee was formed in 1987 after the Metal Mining Division suggested that CIM should have a committee to oversee environmental matters.[403] The Institute envisaged it as an industry-wide forum with representatives from all the technical Divisions, and Carl Anderson, a District Vice President, was appointed its first chairman. The following year, in a wide-ranging review of the Institute and its operations, and the challenges facing Canada's mineral industry, Peter Tarassoff , CIM President in 1990-91, observed that the Mining Association of Canada, the Canadian Petroleum Association, and other groups and their member companies, had adopted environmental codes of practice built on the concept of sustainable development.[404] He proposed that the Institute, whose members would have the responsibility for putting these codes into practice, should also declare a similar commitment. This was accepted by Council in 1990, and the Environmental Committee was charged with preparing a suitable statement.[405] Within the year, a seven-point policy had been drawn up, emphasizing research, education, and the development of new technology and its transfer, as the underpinnings of sustainable development.

CIM's Environmental Policy Statement was formally unveiled at the 93rd Annual General Meeting in Vancouver on April 28, 1991, which also featured a series of technical sessions on the environment.[406] Three years later, during the presidency of Bill Almdal (1994-95), the Environmental Policy Statement was reproduced as a handsome, frameable certificate, and distributed to all CIM members as a reminder of their environmental stewardship.[407]

Whitehorse Mining Initiative

The Institute's concern with the environment and sustainable development had become increasingly evident, and starting in 1991, the

---

[403] CIM, *Council Minutes*, December 10, 1987, minute 13.

[404] P. Tarassoff, "Towards the Year 2000," *Bulletin*, no. 939 (July 1990): 12-20.

[405] CIM, *Council Minutes*, June 11, 1990, minute 16.

[406] *Bulletin*, no. 952 (August 1991): 38-39.

[407] *Bulletin*, no. 1010 (May 1997): 11. Giorgio Massobrio, "Observations by the President," *Bulletin* no. 996 (January 1996}: 6.

Bulletin began to devote special issues to the subject.[408] Another important step was taken in 1994, when under President Bill Almdal, the Institute became a co-signatory of the Whitehorse Mining Initiative (WMI), an accord which brought together all the principal stakeholders in the mining industry, including mining companies and associations, governments, environmental and aboriginal communities, and labour. Among its goals, the WMI called for a sensible balance between protection of the environment, and the mineral exploration and development needed to maintain the mining industry and create new wealth for the Canadian economy. Significantly, the WMI was formally signed during the annual Mines and Energy Ministers Conference, which itself had evolved from an Institute initiative in 1944.

In 1995, under the leadership of President Giorgio Massobrio (1995-96) and Past President David Robertson (1994-95), CIM again became an active environmental partner with industry. The Institute agreed to participate, in a technical capacity, in developing an industry-wide position on the application of the federal government's Mineral Resource Assessments (MRAs) in land-use planning. MRAs had arisen in the 1960s and 1970s as a method of assessing certain areas as potential sources for minerals in the face of the perceived shortages that Patching had focused on in his 1972 presidential address.

The problem was that in the 1990s, the MRAs had become a basis for withdrawing land from future exploration and mineral resource development. The CIM's position was that "The only true 'mineral resource assessment' of any area is the ongoing and open-ended market and knowledge-driven process of mineral exploration itself," and from a standpoint of technical and economic practicability, it was impossible to conclude that a given area had no mineral potential.[409] For example, based on the geological information available in the 1980s, the Lac de Gras region, the site of Canada's first diamond field, and Voisey's Bay, the most significant new mineral discovery in decades, might have been withdrawn from exploration as protected areas.

Mining Hall of Fame

In his after-dinner address at the Engineers Club in Toronto, in 1987, M.R. Brown, then editor of the *Northern Miner*, delivered a light-

[408] Oliver, ibid. Irvine and Gardiner, ibid.

[409] "Towards the Development of an Industry-wide Position on the Use of Mineral Resource Assessments," *Bulletin*, no. 997 (February 1996): 32-33.

hearted talk on "Saints and Sinners I have known." In describing several of Canada's more illustrious prospectors and miners, Brown concluded with a proposal: "Isn't it about time that we in this industry got together and started a Canadian Mining Hall of Fame?"[410] Later, one of the *Northern Miner*'s editorial writers published a story on Brown's suggestion, and the idea was rapidly accepted by the CIM, the Mining Association of Canada (MAC), the Prospectors and Developers Association of Canada (PDAC), the Northern Miner Press, and fifteen of Canada's provincial mining organizations.[411] An ad hoc committee representing the four major sponsors (CIM, MAC, PDAC, and the Northern Miner) met in 1988, and the first annual induction dinner of the Mining Hall of Fame took place at the Royal York Hotel in Toronto on January 21, 1989.

The American National Mining Hall of Fame had held its first induction ceremony only a few months earlier, in September 1988.[412] Its permanent home is in Leadville, Colorado, and features a museum, an archive, and a conference centre. In Canada, the current home of the Canadian Mining Hall of Fame is at Elliot Lake, Ontario.[413] On display in its new Civic Centre, which also contains a small mining museum, are portraits of the individuals who have made outstanding contributions to Canada's mining industry. The criteria for both organizations are similar, but in Canada, living candidates are also considered. Now in its ninth year, the Canadian Mining Hall of Fame includes seven of CIM's past presidents, and fittingly, the Institute's first secretary, B.T.A. Bell.

An Old Name Revisited

In 1981, the *Council Minutes* noted a three-line suggestion that the By-law Revision Committee should consider modifying the Institute's name so that it would contain the same sequence of letters in English as in French.[414] No action was taken in this direction, neither would any be made until December 1987, when the Petroleum Society submitted a formal request to Council for a name change. Tony Anton, Petsoc's Chairman and a future CIM President (1992-93) spoke as follows:

[410] "Mining Hall of Fame Grew from an Off-the-Cuff Remark," *Northern Miner* (July 16, 1990).
[411] CIM's participation was approved by Council on March 25, 1988.
[412] "National Mining Hall of Fame Inducts its First 25 Members," *Mining Engineering* (November 1988): 1009.
[413] "Canadian Mining Hall of Fame Moves to Elliot Lake," *Bulletin*, no. 996 (January 1996): 7.
[414] CIM, *Council Minutes*, "Change of Name - CIM," November 18, 1981, minute 33.

> The Society believes that the petroleum industry is not adequately described by the terms "mining" and or "mineral extraction industry," and that the addition of the word "petroleum" would indicate the true interdisciplinary mix within the Institute.[415]

The idea of adding the word "petroleum" to the Institute's name was certainly not new, having been discussed at length from 1960 to 1964, and then postponed. In support of the renewed proposal, Anton pointed out that the Society represented a significant proportion of CIM's individual and corporate membership, that Petsoc was enjoying steady growth despite the recession, and the Society of Petroleum Engineers (SPE) of AIME was continuing to promote the formation of SPE sections in Canada. After a suitable discussion, Council agreed to form an ad hoc committee to review the proposal, and report back at its next meeting. While some councillors favoured the change, others believed that "no specific discipline should be mentioned in the name of the Institute."[416]

On March 25, 1988, the ad hoc committee, chaired by Bill Almdal, concluded that the addition of the word "Petroleum" would be in the "best interest of the Institute as a national body to publicize its interdisciplinary role."[417] A motion was moved and carried by Council, and the Committee's recommendation appeared in the May issues of the *Bulletin* and *JCPT*. The proposal was then put to a general vote of the CIM membership. The results of the ballot were clear: out of the 4,924 votes cast, 4,256 had agreed to the name change. Pierre Michaud was asked to undertake the necessary steps to petition the Canadian Parliament "to enact a statute to amend the Institute's Charter."[418]

Michaud reported in October 1989 that the preparatory work had been completed, and the petition to amend the Institute's name was before Parliament.[419] CIM President John Laffin, who was also Nova Scotia's Deputy Minister of Mines and Energy, initiated the final actions needed to spearhead the Bill through the House.[420] On June 19, 1990, CIM officially

---

[415] CIM, *Council Minutes*, "Request by the Petroleum Society of CIM for a Name Change of the Institute," December 10, 1987, minute 6.

[416] Ibid.

[417] CIM, *Council Minutes*, "Report from Ad Hoc Advisory Committee on Name Change," March 25, 1988, minute 5.

[418] CIM, *Council Minutes*, "Results of Vote - Name Change," October 17, 1988, minute 4 (c).

[419] CIM, *Council Minutes*, "Name Change," October 25, 1989, minute (E).

[420] John Laffin, interview with author, April 29, 1997.

became the Canadian Institute of Mining, Metallurgy and Petroleum. It was agreed that the official abbreviation would remain the same. In the *JCPT*, CIM President Peter Tarassoff commented:

> CIM's name change is a most welcome and long overdue recognition of the importance of the petroleum industry and its people in the life of the Institute. Congratulations and best wishes to The Petroleum Society of CIM and to all those CIM members associated with the industry![421]

A Road Map for the Institute

Strategic planning is a critical element in the management of a modern organization. Starting in the early 1980s, special committees were investigating everything from finances to reorganization, and Council was becoming more aware with each passing year that CIM needed a road map for its future. The first ten-year plan was introduced in 1982 under the Presidency of Noel Cleland (1982-83). Council was told that the plan:

> attempted to forecast where the Institute was going, and would require modification, from time to time, to recognize new trends and needs, and while the plan set goals, these goals were not cast in stone.[422]

This plan estimated membership growth rates, and probable revenues and expenditures, and projected the expansion of the Secretariat and its staff, publications, and regional and technical meetings. Although the plan was tabled in April 1982, it was not presented to the CIM membership because Council recommended that it should be "sharpened and honed." For the next four years, the ten-year plan would be updated regularly.

By 1986, it was becoming increasingly apparent to Council that the ten-year time span was too long for accurate predictions. In December 1987, Lindsay Milne, President, and René Dufour informed Council that "it had been agreed. . . that a 5-year plan would replace the 10-year plan, and that from now on, the dossier would be the responsibility of the President-Elect."[423] Dufour then reported that all Vice Presidents, and Chairmen of

[421] Peter Tarassoff, "Observations by the President," *JCPT*, vol. 29, no. 5 (September-October 1990): 6.
[422] CIM, *Council Minutes*, "Ten (10) Year Plan," March 19, 1982, minute 40.
[423] CIM, *Council Minutes*, "Report from President-Elect on 5-year Plan," December 1987, minute 7.

Divisions, Societies, and Committees had been asked for inputs by means of a questionnaire, and that the results of these "working groups" would be compiled by March 1988.

The information collected was very valuable because it brought several issues into focus.[424] The 5-year time span was still considered too long, and would be reduced to three years, with the added involvement of the Institute's membership. Planning would be limited to policy matters, while statistical reports on membership, and revenues and expenditures would be prepared by the Secretariat staff. For the first time in Institute history, there would be a concerted effort to design a mission statement with clear objectives, emphasizing members' needs. The organizational structure of CIM would also be examined, along with the redefinition of roles for Council, Divisions, Societies, Committees, Districts, and the Secretariat. The 1988 Three-Year Strategic Plan prepared by Dufour was published in the *Bulletin*.[425] It was implemented when Dufour became President in 1988-89, and carried forward by the succeeding presidents.

A more comprehensive strategic planning process was initiated by President A.D. (Tony) Anton in 1993. A Strategic Planning Committee was established under the chairmanship of Bill Almdal, and an outside facilitator was brought in to provide guidance. In seeking comments on the committee's draft report, Almdal observed:

> The world around us is changing and we must adapt and continue to serve the needs of our members. Change is accelerating and we need to move just as quickly.[426]

After a great deal of time and reflection, and input from a large number of sources, including the many special committees and task forces which had existed since the early 1980s, a Strategic Plan was published in 1994. It set out a Vision and a Mission for the Institute:[427]

> **Vision:** To be The Association of Choice for people in the resources industry.
> **Mission:** To meet the needs of the mining, metallurgy, and petroleum industries by providing quality services and

[424] CIM, *Council Minutes*, "5-Year Strategic Plan," March 25, 1988, minute 7. CIM, *Council Minutes*, "Presentation of 3-Year Strategic Plan," May 1, 1988, minute 7.
[425] René Dufour, "The 1988 Three-Year Strategic Plan," *Bulletin*, no. 917 (September 1988): 11-15.
[426] Bill Almdal, letter soliciting comments on draft planning document, December 16, 1993.
[427] *Strategic Plan: CIM, the Association of Choice* (Montreal: CIM) May 30, 1994,

> value to members through: technical forums, publications, professional and resource networking, promoting our respective industries, student activities, continuation education, and recognition of excellence.

Almdal pointed out that the Strategic Plan "is a living document [that] will attempt to deal with our many challenges, whether they be in the area of publications, membership, or the role of the Institute itself."[428] A permanent Strategic Planning Committee, chaired by the President-Elect, was established in May 1994. Its mandate is to validate and update the Plan on a three-year cycle, and to monitor its implementation.

In referring to the 1997 Strategic Plan, CIM's 1997-98 President, Sandy Laird, confirmed that the fundamental purpose of the Institute was to provide its members with opportunities for knowledge and fellowship. He emphasized that those who had worked on the plan had left 'no stone unturned' in their quest for solutions. Because the 1997 three-year plan would propel the CIM into the next century, it was vital that a firm foundation be built which could withstand an often volatile external environment. He urged that the planning process must never idle, but continue to search for new ways for CIM to provide value to its members.[429]

Throughout its history, the Institute has been devoted to the needs of its members, and the welfare of Canada's mineral industry. From the initial implementation of Branches in 1902, to the first rudimentary technical sections in 1916, the Institute's *raison d'être* has been one of service. Starting in 1947, the Institute began to take a more critical look at how to prepare for the next generation of scientists and engineers, and the burgeoning of technology. In the decades that followed, it diversified and expanded into an organization of more than 12,000 members. As CIM moves into its Centennial year, it brings with it the collective experience of time. If the original Founders could be present today, they would be amazed and proud of what their vision had created.

[428] Bill Almdal, "Observations by the President," *Bulletin*, no. 981 (June 1994): 6.

[429] "1997 Strategic Plan," *Bulletin*, no. 1013 (September 1997): 7. Sandy Laird, "Observations by the President," ibid., 6.

CHAPTER TWO

CIM BRANCHES

Branch meetings have considerable influence, especially in remote camps, in promoting better application not only of subjects strictly technical, but also of problems concerning the improvement of working conditions in mines, and the welfare of the miner's family and home.

A.O. Dufresne, CIM President, 1950-51[430]

The idea of forming Branches was first suggested in 1902 by B.T.A. Bell, esteemed Secretary of the Canadian Mining Institute. He believed that because mining activity was widely scattered across Canada, the majority of members could not attend regular Institute meetings in larger centres, nor could they visit the Institute's office and library. This was at a time when travel was often long and arduous. Bell worried that these members would soon lose enthusiasm for the organization due to limited contact. The answer lay in creating Branches of the Institute in local communities where people with a common interest in the mining industry could meet to discuss technical problems, exchange information, or just relax in a welcoming atmosphere of fellowship. It was never intended for Branches to become autonomous bodies, but rather to bring the Institute to the grassroots of the membership.

The history of CIM Branches tends to parallel that of the mining, metals, and energy industries in Canada. The Branches have always enjoyed the strong support of local operating companies and suppliers, and they have played an important role in the social life of many communities. These communities have often been subject to the boom and bust days typical of the resource industry.[431] For the many people who accepted temporary residence as a way of life, the Branches provided a network. They knew that in the event of a relocation, they could always join another Branch in a new area, or establish one if none existed.

With very few exceptions, the only written record of the Branches is what has been published in the Institute's *Bulletin*.

---

[430] A.O. Dufresne, address to Red Lake Branch, District 4, October 6, 1950, typescript, vol. 23, NAC.

[431] Rex A. Lucas, *Minetown, Milltown, Railtown: Life in Canadian Communities of Single Industry* (Toronto: University of Toronto Press, 1971) p. 96-99.

The Earliest Branches

CIM's first Branches were formed in 1902 at Sherbrooke, Quebec; Kingston, Ontario; and, Nelson, British Columbia. Their story illustrates how frequently Branch names and jurisdictions have changed, and how Branch activity has often reflected the local mining scene. The branch at Sherbrooke was originally called the Eastern Townships Branch, and its first Chairman was George R. Smith, an Institute Founder, and President from 1905 to 1907. It suffered a membership decline soon after its establishment, and was reorganized in 1910 by John J. Penhale, another Institute Founder, who became its Chairman. It folded by 1923 when copper mining in the region declined. A revival of copper mining led to the formation of a new Sherbrooke Branch in 1952.[432] Then, in 1964, after a period of inactivity, the Branch transferred its moveable assets to Headquarters. Its final demise, like that of so many other Branches, was linked to the cessation of mining activity in the region.

The branch at Kingston was closely associated with the School of Mining and Agriculture at Queen's University, in operation since 1893. It was formed as the Eastern Ontario Section, and renamed the Kingston Branch in 1909. In 1919, Kingston was included in the Hastings District Branch, which also encompassed Deloro, Marmora, Madoc, Tweed, Sulphide, Belleville, Fenton, Havelock, and Cordova Mines. This region saw a lot of different mining activity over the years, from iron ore at Marmora, to fluorspar at Madoc, and gold at Deloro, later the site of an important refinery which treated cobalt-silver ores from Cobalt. By 1934, the Hastings District Branch had folded. When a new Kingston Branch was formed in 1980, it was supported by members from Queen's University and the metals industry in Kingston.

The history of the British Columbia Branch at Nelson is even more complex. It was established on September 10, 1902, during the Institute's summer excursion to British Columbia, and was envisaged as a "branch for the Province" rather than as a local Branch. Its first Chairman was S.S. Fowler (1900-01 CIM President), and its Secretary-Treasurer was Selwyn G. Blaylock (1934-35 President). Its name was changed to the Western Branch in 1908. By this time it had a membership of 150, and Frederick Keefer (1907-08 President) was its Chairman. He was succeeded the

[432] "Sherbrooke Branch. One of the Institute's Earliest Branches Again Active," *Bulletin*, no. 480 (April 1952): 240-241.

next year by Fowler, who served for an additional two years. The Western Branch met at Trail instead of Nelson, perhaps reflecting the growing importance of the Consolidated Mining and Smelting Company (Cominco) smelter in the region.

The Western Branch encompassed members in British Columbia, Yukon Territory, Alberta, and interestingly, the State of Washington. As the apparent western arm of the Institute, it opposed the formation of other branches in B.C. and Alberta. Its regional jurisdiction was affirmed when it was renamed the British Columbia Division in 1919, with Charles Camsell (1947-48 President) as Chairman. Included were Branches at Vancouver, Nanaimo, Anyox, and Nelson. H. Mortimer-Lamb, the Institute's Secretary-Treasurer Emeritus, served as the Division's Secretary from 1920 to 1945, and occupied an office at 902 Tower Building in Vancouver. When the Institute formed the first of its new technical divisions in 1945, the British Columbia Division was renamed the B.C. Section. Until the advent of District meetings, the British Columbia Division and its successor, the B.C. Section, organized Annual Western Meetings of the Institute. In 1977 the executives of District 6 and the B.C. Section were combined, and in 1980 the B.C. Section was amalgamated with the Vancouver Branch.[433]

The Institute's oldest, still active Branch was established at Cobalt in 1907. Its first Chairman was Arthur A. Cole, a future Institute President (1916-18). A very active Branch in its early years, it was disbanded in the 1930s after silver prices dropped to thirty cents an ounce. All of the mines in the area closed except for the O'Brien, which had been kept going by laid-off employees who had leased the property from the company. As the future of Cobalt seemed destined to be that of a ghost town, there was growing demand for cobalt for use in alloys such as stellite, and later as an isotope (Cobalt 60) in medicine. By the end of World War II, silver and cobalt had become equally desirable metals, and mining was again active in Cobalt.[434] This led to the revival of the Cobalt Branch in 1949.

The Porcupine Branch at Timmins was one of the luckier ones. It has been in continuous operation since its founding in 1911. Following the discovery of gold in 1908, the Hollinger, McIntyre, and the Dome mines began operations in 1912. By the 1940s, there were eleven mines in the area. The Porcupine Branch had between 400 and 500 members, a very active technical program, and its annual meetings were well attend-

[433] CIM, *Council Minutes*, "Combine District No. 6 and B.C. Section," April 21, 1977, minute 67.

[434] L. Carson Brown, "Cobalt - The Town with a Silver Lining," *Canadian Geographic* (July 1963): 3-12.

ed.[435] Its future was assured by the discovery of the Kidd Creek copper-zinc orebody in 1964.[436] Production began within two years, and an electrolytic zinc plant commenced operations in 1972, followed by a copper smelter and refinery in 1981. Falconbridge, now part of Noranda, acquired Kidd Creek Mines in 1986. The Porcupine Branch continues to be well-supported by the mining industry.

Branch Diversity

The focus of Branches reflects the interests of their members, which in turn depends on the sector in which they are employed. Some Branches have become specialized as a result. The Hamilton Branch is *de facto* an iron and steel branch, but is not associated with the Metallurgical Society. Similarly, the Calgary Branch is mainly composed of petroleum people, but is separate from the Petroleum Society. The Crowsnest Branch is a coal branch, and the Oil Sands Branch represents a unique sector of the energy industry. The Ottawa Branch draws most of its members from federal government establishments such as CANMET and the Geological Survey of Canada. The Branches in Toronto and Vancouver might be described as corporate office branches, as those cities contain the head offices of many major and junior mining companies, as well as mining consultants, lawyers, stock brokers, and bankers.

A Fleeting Existence

Of the more than 100 Branches that have existed during the Institute's one hundred years, some have had remarkably short lives due to the fortunes of the mining or metallurgical activities with which they were associated. The North of '53 Branch, organized in 1929 at The Pas, Manitoba, disbanded in 1934, and its members were encouraged to join the Flin Flon Branch. The Atikokan-Steep Rock Branch, formed in 1959, was discontinued in 1964 after mining ceased in the area. The Taku Branch, formed at Tulsequah, B.C. in 1955, folded within three years, and the Drumheller Branch only lasted from 1939 to 1942.

As some Branches folded and new ones sought approval, Council began to question whether or not there were too many. When a Whitehorse

[435] C. Bruce Ross, interview with author, February 7, 1996, Toronto.
[436] L. Carson Brown, "The Golden Porcupine," *Canadian Geographic* (January 1967): 5-17.

Branch was proposed in 1955, Council decided that there were not enough supporting members and initially refused the request. The matter was referred to the Executive Committee which proposed extending the jurisdiction of the Mayo Branch to include Whitehorse and Cassiar, and renaming it the Yukon Branch. A spokesman for the Whitehorse group expressed disappointment that it had been turned down, and pointed to the impracticability of their attending Mayo Branch meetings. He also emphasized that potential Institute members had been discouraged by the complete lack of local activities. After reconsideration, Council approved the formation of the Whitehorse Branch. Within three years it had became inactive due to industry conditions. It was reactivated in 1970 as the Yukon Branch, resumed its original name in 1977, and is still active.

"Unofficial" Branches

There were two instances of "unofficial" Branches in the Institute's history. The Flanders Branch established in Belgium during World War I under the leadership of Colonel John J. Penhale was the more serious of the two. Composed of about twenty like-minded and energetic men, it promoted an atmosphere of support and camaraderie at the Front.

The Balkan Branch in Yugoslavia was formed in jest. In a photo published in the *Bulletin* in 1936, its members, R.F. Lethbridge, Major George C. Riley, and F.E. Tracey and four Yugoslav gentlemen, are shown with several horses tethered nearby. They offer their cordial greetings to the Ottawa Branch and "By standing and silent vote the members are reverently adopting a resolution of profound condolence with themselves at missing the meeting in Canada."[437]

Although in four years time Canada would engage in another world war, there is no record of any CIM members having formed another "unofficial" Branch in either Great Britain or on the Continent.

Branches and CIM Presidents

Many future CIM presidents began their Institute involvement at the Branch level. In addition to those already mentioned in the story of the British Columbia Division, the B.C. Section had three other chairmen who became well-known at the national level: Desmond F. Kidd

[437] "The Balkan Branch, C.I.M.M.," *Bulletin*, no. 291 (July 1936): 263.

(1952-53), John F. Walker (1956-57) and E.H. Caldwell (1979-80). Kidd and Walker had also been chairmen of the Vancouver Branch, as were H.M. Wright (1970-71) and W.E. Stanley (1991-92). In Ontario, E. Coste (1903-05), D.E.G. Schmitt (1980-81), C.B. Ross (1986-87) and D.S. Robinson (1993-94) had all served as chairmen of the Toronto Branch. Four prominent members of the Quebec City Branch also became Institute leaders: A.O. Dufresne (1950-51), Paul E. Auger (1968-69), Jean-Paul Drolet (1979-80), and Yves Harvey (1998-99). No fewer than five Presidents served as chairmen of the Montreal Branch: J.G. Ross (1935-36), W.G. McBride (1941-42), J.H. Schloen (1972-73), R.H. Gray (1982-83), and R. Dufour (1988-89).

Often an executive position in one Branch led to involvement in another. In 1947, C. Bruce Ross, then a young man of 27, joined the Porcupine Branch. He volunteered to serve on a Branch committee, and within two years became its chairman. After he moved to Toronto in 1961, he continued his Branch involvement becoming Chairman of the Toronto Branch.

Still others would first lead the Mining Society of Nova Scotia before taking on the CIM Presidency, among them John E. Hardman (1898-1900), Donald H. McDougall (1919-20), Thomas J. Brown (1925-26), Michael Dwyer (1937-38) and John J. Laffin (1989-90), who was also a chairman of Halifax Branch, formed in 1936.

In some cases, men became active in Branch activities after finishing their term as president. Arthur A. Cole (1916-18), was the first Chairman of the Cobalt Branch, and a director for several more years.

Branch Governance

As the Branch concept expanded, the Institute realized the need for a common set of regulations governing their operation. On October 6, 1905, an Institute committee determined that "a Branch shall not consist of less than 10 members in good standing in any province, territory, district, town or mining camp." The phrase "in good standing" meant that Branch members had to pay regular dues. The rules also stated that Branches "may discuss and finally deal with any matters of purely local interest." When some Branches signalled their intention to lobby provincial governments, this was amended in April 1906 to read "branches may discuss any matters brought before them, but no action shall be taken, and their decisions shall in no way affect or bind the general body of the Institute."

In the early years, the Branches were the main subdivisions of the Institute. Accordingly, they played a more prominent role in the governance of the Institute than they do today. Until a provincial district structure was implemented in 1922, Branches had direct representation on Council. In 1910, Branches with more than 40 members were also given direct representation on the Nominating Committee for Institute officers.

Branch Membership

Participation in the Branches has traditionally been open to anyone with an interest in the minerals industry. Because they did not have to meet the same membership criteria as for the Institute, Branch members did not need to be 'national' members unless they served as representatives on Council. In 1937, Council agreed "That the policy of admitting non-members of the Institute into membership in local branches be left in the hands of the local branches to decide individually."[438] This meant that in some Branches, people such as the local bank manager were welcome to join. It has also meant that some Branch members who were eligible to become Institute members have not seen the need to do so. The issue of national versus Branch membership, and the benefits of belonging at both levels, have been debated many times over the years in the pages of the *Bulletin* and in meetings of Council.

Branch Meetings and the Exchange of Knowledge

In the early days, Branch meetings were considered an alternative to the Institute's Annual General Meeting as a venue for the presentation of technical papers. Many of these were later published in the *Bulletin*. Papers presented during the first years of the Cobalt Branch, established in 1908, included "Sampling of Cobalt Ores at Copper Cliff Smelter," by Arthur A. Cole, "The Application of Electricity to the Location of Ore Bodies," by Alfred Williams, and "The Mispickel Deposits at Arsenic Lake," by Capt. J E. Leckie. In 1910, "A largely attended meeting of the Cobalt Branch was held in July, at which Professor Robert H. Richards, the distinguished authority on ore-dressing, delivered a lecture on the subject, having special reference to the conditions obtaining in the Cobalt District." At this time, the Branch was attracting over 100 members to its meetings. It had gener-

[438] CIM, *Council Minutes*, "Conference of Branch Executives," 1937, p. 78, vol. 5, NAC.

ated so much interest in the community that it had been given a room in the Town Hall for a library and mineral exhibit.

The Branches were very important for the transfer of knowledge and experience in more remote regions, and the presence of Institute men who were experts in their fields undoubtedly resulted in improvements in the sometimes haphazard practices of the day. Frank Loring, an Institute Founder, was among the earliest to engage in work in the Cobalt district in 1904. In a paper on mining in Cobalt presented at an Ottawa meeting in 1908, Loring stated that during the first two years of the Cobalt discovery, the only object had been to extract the silver ore as fast and as cheaply as possible. This was often done by using extravagant mining and sorting methods with the result that:

> many thousands dollars worth of silver lie buried in dumps, often covered with waste, which, had more economic methods been adopted, would have largely increased the output of the district.[439]

Because many mines were operated by inexperienced men, simply put, "If the silver couldn't be recognized, then the ore was not saved." The problem was compounded by the fact that there were over one hundred individual mines in the area. What Loring felt was needed was a "miner-like scientific manner," where methods could be openly discussed, and improvements made.

Throughout the years, presentations and discussions at Branch meetings have often revolved around timely topics. In 1954, members of the Winnipeg Branch were eager to hear about the uranium mining boom, and nuclear technology. The program included a talk on "Radioactivity Surveys with the Airborne Scintillometer," a film on helicopter operations in the Beaverlodge area, and a film entitled "A Picture of the Atomic Age." During the same year, the Flin Flon Branch welcomed R.J. Banata of the Minerec Corporation, New York, and W. de Villiers of the Rhokana Corporation, Northern Rhodesia. Banata spoke on flotation practices in larger mines in North America, and de Villiers elaborated on operational problems at Rhokana, and then entertained the members with descriptions of Northern Rhodesia and parts of Africa. Such speakers brought an international flavour to Branch meetings.

[439] Frank C. Loring, "Mining at Cobalt," *Bulletin*, no. 1 (February 1908): 69-73.

Some twenty years later, there were new concerns. In January 1973, the Thetford Mines Branch assembled to hear a talk by W.H. Gauvin, Director of Research and Development at Noranda Mines Limited, on the future of the Canadian metallurgical industry. He projected how the rise in population, the Gross National Product, capital investments, taxation revisions, and pollution might affect primary and secondary industries in the future. At the Gaspé Branch in December, members watched a film on the conservation of wildlife, courtesy of Canadian Industries Limited, which also provided oysters for the members' enjoyment.

The Lively Side of Branch Meetings

Branch meetings could also be lively and entertaining. One month after it was formed in 1908, the Montreal Branch held a dinner meeting at the Engineer's Club. The Chairman, George E. Drummond, a Charter Member of the Institute, and the Vice-Chairman, John Hardman, had invited two well-known guest speakers: Willet G. Miller, the Provincial Geologist of Ontario and Institute President in 1908-10, and Reginald W. Brock, then Director of the Geological Survey of Canada. As was common at the time, the evening was interspersed with singing, and toasts to "the Manufacturing Interests, Financial Institutions, the Western Branch of the C.M.I., the Mining Industry of Australia, McGill University, and the Secretary of the Institute."[440] Support for the Montreal Branch came from university people, the city's metal industries, such as foundries, and mining company head offices.

The goal of the Sudbury Branch was to be "the liveliest Branch in Canada."[441] Established in 1911, it often met at the Sudbury High School. Its first Chairman, Robert G. Leckie, had been President of both the Mining Society of Nova Scotia, and the Federated Canadian Mining Institute. Another distinguished member, John A. Dresser (1923-24), had also been a prime mover in the Branch's formation. Located in one of Canada's oldest mining districts, the Branch has been active for more than eighty-five years.

In more modern times, Branch meetings and activities have become more varied. At the Thetford Mines Branch, members would go home after work, have supper, then drive to the monthly meeting which was usually held at the golf club. After refreshments and fellowship, a talk would

[440] "Montreal Branch Meeting," *Bulletin*, no. 2 (May 1908): 72.
[441] "Progress of the Canadian Mining Institute," *Bulletin*, no. 91 (November 1919): 1114.

be given, and the evening would be topped with a 'night-cap.' Some Branches like those in Vancouver and Winnipeg held weekly luncheon meetings, with or without speakers, which gave local and visiting members an opportunity for informal contact. Many Branches organize safety nights, ladies' nights and other special events.

The Annual Meeting of the Newfoundland Branch brought together people from different parts of the province, such as Labrador City and Baie Verte. Aside from the presentation of technical papers, the program often featured a few sheets of curling, and for anyone who was a newcomer to the province, induction into the Screecher's Club. This was a Newfoundland tradition involving sitting in a dory, downing a glass of rum, and kissing the cod. Annual golf tournaments have become a feature of the Montreal and other Branches. Branch meetings continue to be memorable events that tie the mineral community together, promote fellowship, and lead to long-lasting friendships and professional associations.

Distinguished Lecturers

Since its establishment in 1967, the Distinguished Lecturers program has brought a steady stream of highly qualified speakers to Branches across Canada. The program is supported by the Canadian Mining, Metallurgical, and Petroleum Foundation, and as many as eight Institute members are named CIM Distinguished Lecturers each year. Lecturers are chosen on the basis of distinguished service and achievement in science, technology, management or education related to the minerals industries. They have provided Branch members with new insights, and their talks would often engage the members in lively discussions. They have also helped to bridge the boundaries between different sectors of industry. For example, in 1973, Branch members from the Val d'Or gold mining district heard L.C. Piuze, Vice President and General Manager of Lake Asbestos of Quebec Limited, speak on "The Influence of the Asbestos Industry on the Canadian Economy and Control of Air Pollution in the Quebec Asbestos Mines."

Sometimes students and the interested public were invited to attend Branch meetings to hear Distinguished Lecturers. In 1981-82, T. Alan O'Hara of the Hudson Bay Mining and Smelting Company gave thirty-four lectures with students in attendance at Branches from Schefferville to Victoria. Typical of the support provided by industry, Hudson Bay covered much of the expense. For Branches, the benefits of the kind of continuing education inherent in the Distinguished Lecturers program are invaluable.

Mine Rescue and Mine Safety

One of the more serious aspects of Branch activities were competitions in mine rescue. The Rocky Mountain Branch, formed at Lethbridge in 1912, was the first to inaugurate mine rescue and first aid competitions in 1916. This soon evolved into an annual event for all Branches in Alberta. Since the horrendous tragedies of the Frank Slide in 1903, Bellevue mines in 1910, and Hillcrest in 1914, most people were acutely aware of the dangers faced by coal miners in the Crowsnest Pass region.[442] William R. Wilson (1922-23 President), the General Manager of Crows Nest Pass Coal Company, and the Chairman of the Rocky Mountain Branch, responded to a paper given by Duncan McDonald on "Mine Rescue and First Aid Work" and his suggestion of holding competitions, by stating that:

> he was prepared to offer a cup as First Prize and suggested that a committee be formed to organize some plans whereby the wishes of the members in the different districts could be met regarding these competitions. . . a portion of the funds of the Branch could [also] be used for the purpose of providing 2nd, 3rd, and 4th prizes.[443]

After a large number of fatal accidents occurred in 1916, more than in any other previous year, coal mining companies in the region decided to train miners in rescue techniques. This was technology transfer of a very different sort. Metal mines also joined in. The Centre Star and War Eagle mines gave resuscitation demonstrations, using pulmonators, to the St. John's Ambulance Association. In 1917, mine rescue cars were installed in districts where the mines were in fairly close proximity, and in the more remote regions, designated railway stations were provided with rail cars to accommodate mine rescue teams and their equipment.

Yearly competitions in mine rescue were organized by the Rocky Mountain Branch. The organizing committee included Duncan McDonald, and Lewis Stockett, General Superintendent of the Coal Mine Branch of the Canadian Pacific Railways Department of Natural Resources, and a future CIM President (1926-27). In 1921, the province was divided into northern and southern regions for the competition event, with elimination

---

[442] The Hillcrest mine remains Canada's worst underground disaster, killing 189 men out of the 235 on shift on June 19, 1914.

[443] Rocky Mountain Branch, Minutes, August 12, 1915, vol. 12, NAC.

rounds, and finals. One result of these "games" was that the Vice-Chairman of the Southern region, Robert Strachan of Fernie, B.C., was appointed by the government to study mine rescue apparatus and recommend a standard appliance for use across Canada.[444] The subject of mine safety finds a modern expression in the John T. Ryan Trophies.

Ryan was a mining engineer from Penn State University who was in charge of mine safety, rescue, and appliance testing for the U.S. Bureau of Mines before co-founding the Mine Safety Appliances Company. In 1915, he had worked with Thomas A. Edison on the design and development of the Edison Safety Cap Lamp, which became widely used in coal and metal mines. Ryan was a member of CIM, AIME, several coal mining associations in the U.S., and a Charter member of the National Mine Rescue Association. Under his direction, the Mine Safety Appliances Company and its Canadian affiliate became the largest manufacturer of mining and industrial safety equipment in the world. In 1937, the company sponsored two Accident Prevention Prizes for the best papers on the subject presented at an Institute meeting. After Ryan died in 1941, the John T. Ryan Trophies were established in his memory, and presented to mining companies with the lowest accident frequencies. In a report published in the *Bulletin* in 1954, the Committee overseeing the nominations reported that the competition was a powerful tool in reducing accident frequencies, adding to the efficiency of operations, and ultimately, to the contentment of miners and their families. Although the Canada Trophies continue to be presented at the Institute's Annual General Meeting, the six Regional Trophies are given out at District or Branch meetings, which brings the competitions back full circle to the initiative of the Rocky Mountain Branch.

Branch Awards

Of the Institute's many awards, the District Distinguished Service Awards were established to especially recognize significant technical, professional or managerial contributions at the local or Branch level. The Mel W. Bartley Award is presented annually to the Branch judged to have made the most progess toward achieving the aims and objectives of the Institute. It is named after the 1976-77 CIM President and was established in 1977 by the Thunder Bay Branch. The award was originally presented to District 4 branches only, until 1991.

[444] "Rocky Mountain Branch," *Bulletin*, no. 107 (March 1921): 246-248.

The New Brunswick Branch launched the Dr. W. J. Wright Award in 1979. William J. Wright was born in 1881, in Bear River, Nova Scotia. He earned a BSc at Acadia University, then obtained a BA, an MA, and a PhD in geology at Yale University. He served with the Canadian Infantry from 1916-18, and after the war, worked for the Geological Survey of Canada in the Maritimes. After spending several years in the U.S. with the Whitehall Petroleum Corporation, and Standard Oil in New Jersey, he was appointed Provincial Geologist for New Brunswick in 1928. During this time, he carried out a detailed study on the Austin Brook iron deposits, which led to the recognition of lead and zinc in association with the iron formation, and eventually to the discovery of the famous Brunswick No.6 and No.12 deposits. Wright was a longtime member of CIM. The award in his honour is presented annually to persons who have made outstanding contributions to the New Brunswick mineral industry. Fittingly, Wright was the first recipient of the award; it was presented to him on his ninety-ninth birthday in 1980.

Presidential Visits and Field Trips

Whenever a CIM President visited the Branches, he was frequently treated to a special field trip. Sometimes, a Branch would try especially hard to impress an Institute officer, leading Jean-Paul Drolet (President in 1978-79) to comment, "Wherever you go, they try to kill the president."[445] Apparently, he had experienced everything from rustic rides underground to hazardous excursions across the countryside, courtesy of over-zealous Branch members. Drolet was a very experienced mining man, having been involved in the exploration and development of the iron ore deposits at Gagnon, Quebec. He was not easily shaken, but some field trips clearly exceeded his expectations.

CIM Executive Director, Gordon Tapp, also recalled a field trip while visiting the Baie Verte Branch in Newfoundland. The group had flown to St. John's, then had taken the famous "Newfie Bullet" train to Badger. "It was my first trip to Newfoundland and the first thing that impressed me was that there was more beer on the train than people." From Badger the group travelled by boat to Tilt Cove, and somehow, "all the beer and other refreshments got on one of the other boats. However, we survived and made good and proper plans for the trip back."[446]

[445] Jean-Paul Drolet, interview with author, July 18, 1996, Ottawa.
[446] "Personality Profile. E. Gordon Tapp," *Bulletin*, no. 809 (September 1979): 50-58.

During his presidential term, Jean-Paul Drolet was able to visit more than twenty Branches. An enthusiastic and captivating speaker, he often focussed on Canadian mineral development in a national and international context. As Senior Assistant Deputy Minister (International Minerals) of the Department of Energy, Mines, and Resources, his perspective was unique. Drolet was the only President to preprint his addresses, which he later assembled into a bound book and deposited in the Institute's archives. The book is a fascinating, late 1970s view of such topics as "Mining and Conservation," "The CIM and its Role in Canadian Affairs," "Minerals and the Law of the Sea," and "Mineraux et Autres Questions Politiques."[447]

If Drolet had included a dedication in the book, it would have been to the students who would eventually assume responsibility for managing Canada's mineral resources. But his interest in education did not stop with university students. While making his round of Branches, he frequently made detours into local grade schools to speak informally with younger students about miners and the importance of the mining industry.[448] Often carting maps, diagrams, and ore samples in his briefcase, he was an Institute ambassador par excellence.

Student Support

CIM Branches have been especially active in supporting students, and many examples can be found in the Branch reports in the *Bulletin*. Starting in 1921, the Northern Alberta Branch awarded prizes for the best student papers on original research work in mining and geology. The Flin Flon Branch organized a series of six lectures during the summer of 1945 for university students employed by the Hudson Bay Mining and Smelting Company.[449] In 1960, the Hamilton Branch inaugurated a student loan fund for students registered in metallurgy at McMaster University and the Ryerson Institute, and in 1974, the Gaspé Branch awarded $1,000 in bursaries to students living in the Gaspésie region. The Edmonton Branch offers two annual scholarships of $1,250 to mining students at the University of Alberta. A number of other Branches have also established scholarship programs. Branches in areas where there are colleges and universities host special student's nights, and subsidize their attendance at regular dinner

[447] Jean-Paul Drolet, Addresses Presented to the Members of the Canadian Institute of Mining and Metallurgy, Jean-Paul Drolet, President 1978-79, Montreal.
[448] Drolet, interview, *ibid*.
[449] "Fine Service to Engineering Students," *Bulletin*, no. 399 (July 1945): 307-308.

meetings. For many students, Branches have been their first and most important direct contact with the Institute.

In the 1980s, as in other periods in the past, there was mounting concern about the lack of students heading for occupations in the minerals industry. This "career crisis" prompted CIM President Peter Tarassoff (1990-91) to recommend that Council establish a Careers Task Force to examine what could be done. Chaired by Bill Stanley (1991-92), the fifteen-member Task Force produced a position paper called "Attracting Students to the Mining, Minerals, Metallurgical & Petroleum Industries." The paper focussed on strategies to improve the industry's image, initiatives to educate teachers and parents of four student age groups, and the creation of summer employment opportunities. Above all, the Task Force emphasized that:

> We the industry must get our message out to students, teachers and the general public. . . since children are our future. . . Their perception of the minerals industry will shape government policy, and, consequently, the long term health of our industry in this country.[450]

Most children are intrigued by science in the third grade, and tend to make career decisions between grades eight and ten. Because many of the Task Force's strategies hinged on effective liaison between the local minerals industry and elementary and high schools, Branches were seen to provide the important link in the communication process. As many Branch members were parents, they could also influence education and career choices.

For many years, Branches had already been active in promoting the industry at the local school level. The New Brunswick Branch, for example, prepared and distributed sets of rocks and minerals, and the Sudbury and other Branches had become involved with local Science Fairs.

The contributions of Canadian women's mining associations to education must also be mentioned. They include the Women's Association of the Mining Industry of Canada (WAMIC), the Women's Association of the Montreal Branch of CIM, the Greater Vancouver Mining Women's Association, and the Women's Auxiliary of the Halifax Branch of the Mining

[450] CIM Careers Task Force, "Attracting Students to the Mining, Minerals, Metallurgical & Petroleum Industries," position paper, October 1991 (Montreal: CIM) p. 44.

Society of Nova Scotia. The Montreal Branch Women's Association has for a long time awarded annual student scholarships, and the WAMIC has provided student bursaries since 1940, student loans since 1952, and has supported the earth sciences departments of universities and colleges across the country. For many women scientists and engineers who have embarked on careers in the minerals industry, the concern shown by these associations has made all the difference.[451]

Affiliated Student Associations

When two CIM'ers engage in conversation, the question is often asked, "So how did you become interested in CIM?" More often than not the answer is, "I got involved at a student branch"; or "I was invited to Student's Night"; or "I was awarded a CIM scholarship which helped me through school."[452]

Although not officially classified as Branches, student associations affiliated with the Institute have existed since the very early years. The first were the Mining and Metallurgy Society (1907), and the Adams Club, both at McGill University. Also in Montreal, the Institut Minière à l'École Polytechnique Montréal, now called Conseil étudiant (mines), has been in operation since 1909. At Queen's University, the Mining and Metallurgical Society (1908) had the largest student membership of its time.

Other groups followed, such as the Geological Society at the University of Saskatchewan, the Geology Club at the University of Manitoba, the Mining and Metallurgical Club and the Rocky Fellers Club at the University of Toronto, the Dawson Club at Dalhousie University, the Mining Society of Nova Scotia Technical College, the G.M. Dawson Club at the University of British Columbia, formed in 1929, and the Mining and Geological Society at the University of Alberta.

For many students, joining a Branch after graduation has been their next CIM involvement. Participation in the Branches by the younger members has always been welcome, and has encouraged them to become proficient in writing papers, public speaking, organization, committee work, and in conducting business meetings. Even though most engineers and scien-

---

[451] "The Women's Association of the Mining Industry of Canada: Commitments to the Community over Seventy-Five Years," typescript (Toronto: WAMIC) 2 p. Marjorie Sheridan, "The Women's Association of the Mining Industry of Canada, Toronto," paper, 7 p. Helen Fitzer and Mary Keen, "The Greater Vancouver Mining Women's Association, History to 1992," typescript, 3 p.

[452] Don Worth, "Observations by the President," *Bulletin*, no. 1007 (February 1997): 6.

tists would have learned some of these skills while in university, the Branch experience enhances their education.

CIM Branches have played an important role throughout the Institute's history. Their story will continue to evolve as new frontiers open to mineral development, and older regions become dormant. As the Institute approaches the end of its first century, a new branch, the North Central B.C. Branch, has been proposed.[453] Another has been planned for Santiago, Chile, which would have the distinction of being CIM's first official branch outside Canada. Other Branches will likely be formed in places like Voisey's Bay, Labrador, and Lac de Gras, Northwest Territories. What binds them together are the common goals of the exchange of knowledge, the improvement of the minerals industry, the welfare of the community, and fellowship in CIM. A.O. Dufresne's words are just as true today as they were fifty years ago.

[453] CIM, *Council Minutes*, "Request from B.C. Branch," August 17, 1997, minute 13.0.

CIM BRANCHES — PAST AND PRESENT

Name	Remarks
BRITISH COLUMBIA	
Bridge River Branch	Established 1937 at Bralorne. First chairman: H.L. Hill. Disbanded in 1944.
Britannia Beach Branch	Established with Council approval October 3, 1949. First chairman: W.G. Hatch. Discontinued in 1958.
British Columbia Branch	Established September 10, 1902 at Nelson, as the western arm of the Institute. First chairman: S.S. Fowler (CIM President, 1900-01). Name changed to Western Branch on January 15, 1908.
British Columbia Division	Succeeded the Western Branch in 1919. First chairman: Charles Camsell (CIM President in 1947-48). Consisted of the Branches at Vancouver, Nanaimo, Anyox and Nelson. Renamed British Columbia Section on October 5, 1945. Amalgamated with Vancouver Branch on February 15, 1980. H. Mortimer-Lamb (CIM Secretary Emeritus) was the Division's secretary from 1920 to 1945.
British Columbia Section	See British Columbia Division.
Crowsnest Branch	Established June 14, 1972 to serve members in the Fernie-Sparwood area. Approved by Council on March 3, 1972. First chairman: J.R. McIntosh. Still active. There was a Crowsnest Section of the Rocky Mountain Branch at Blairmore, Alberta in 1934.
Georgia Strait Branch	Established at Campbell River with Council approval on March 20, 1989. First chairman: M. Petrina. Still active.
Granisle Branch	See Omineca Branch.
Kootenay-Boundary Branch	Established at Nelson in 1919.

Name	Remarks
Merritt Branch	Established on February 22, 1963. First Chairman: A.M. Laird (CIM President in 1997-98). Dissolved by Council on April 29, 1976. A Merritt and District Branch was originally formed in 1923.
Nanaimo Coal Branch	See Vancouver Island Branch.
Nelson Branch	Established in 1919. Re-established in 1951 with H.E. Doelle (CIM Vice President in 1954-55) as chairman. Discontinued in 1972. Nelson was originally part of the British Columbia Branch formed there in 1902.
North Central British Columbia Branch	Establishment approved in principle by Council on August 17, 1997.
North Coast Branch	Established at Anyox on September 20, 1919. First chairman: A.W. Davis. A North Coast Local Branch was formed at Prince Rupert in the same year. Ceased to exist by 1933.
Northern British Columbia Branch	Established at Prince George on March 4, 1966. First Chairman: W.G. Clark. Discontinued in 1983.
Northwestern British Columbia Branch	Established at Prince Rupert with Council approval on April 25, 1974. First chairman: C.L. Stafford. Discontinued in 1983.
Omenica Branch.	Established as the Granisle Branch with Council approval on March 15, 1974. First chairman: J.J. Kalmet. Name changed to Omineca Branch with Council approval on March 19, 1982. Chairman: T. Schroeter. Based at Houston, later at Smithers. Inactive since 1991.
Princeton Branch	Originally established as Similkameen Branch in 1949. First chairman: F.A. McGonigle. Re-established with Council approval on October 4, 1984. Chairman: Don Rankin Amalgamated with South Central British Columbia Branch on February 16, 1991.

Name	Remarks
Rossland Branch	Established on May 25, 1921. First chairman: M.E, Purcell (CIM Vice President in 1917-18). Dissolved in 1924.
Similkameen Branch	See Princeton Branch.
South Central British Columbia Branch	Established at Kelowna with Council approval on March 21, 1969. Inaugural meeting on June 14, 1969. First chairman: T.W. Waterland. Absorbed Princeton Branch in 1991. Still active.
Stewart Branch	Established on June 14, 1926. First chairman: P.E. Peterson. Discontinued by 1927. Briefly reactivated from 1974 to 1975.
Sullivan Branch	Established at Kimberley on February 14, 1946. First chairman: E.G. Tapp (CIM Executive Director, 1965-79). Successor to Kimberley Section of Rocky Mountain Branch. Still active.
Taku Branch	Established at Tulsequah on October 13, 1955. Disbanded in 1958.
Trail Branch	Established as the West Kootenay Branch on October 24, 1945. Approved by Council on December 13, 1945. First chairman: R.J. Armstrong. Name changed to Trail and District Branch in 1952, and Trail Branch in 1993. Still active.
Trail and District Branch	See Trail Branch.
Vancouver Branch	Established in 1919. First Chairman: R.W. Brock. Still active.
Vancouver Island Branch	Established as the Nanaimo Coal Branch in 1919. Reorganized as the Vancouver Island Branch on June 2, 1921. First chairman: George O'Brian. Affiliated with Vancouver Branch. Dissolved in March 1935.
Victoria Branch	Established in 1949. First Chairman: S.S. Holland. Still active.

Name	Remarks
Western Branch	Established at Nelson on January 15, 1908, as successor to the British Columbia Branch. First chairman: A.B.W. Hodges (CIM Vice President in 1910-11). Encompassed Institute members in British Columbia, Yukon Territory, Alberta and State of Washington. Name changed to British Columbia Division in 1919.
West Kootenay Branch	See Trail Branch.
Wolverine Branch	Established at Tumbler Ridge on September 29, 1983. First chairmen: J. Balmer, M.Petrina. Inactive since 1997.
YUKON TERRITORY	
Faro Branch	Established in 1977. First chairman: H. Hong. Inactive from 1983 to 1987. Reactivated with Council approval on December 12, 1987. Chairman: J.F. Huntley. Inactive since 1992.
Mayo Branch	Established in April 1949. First chairman: H. Brodie Hick. Reactivated on December 19, 1959. Discontinued in 1972.
Whitehorse Branch	Established 1955 and operated to 1959. Reactivated in 1970 as Yukon Branch with A.C. Ogilvy as chairman. Approved by Council on January 25, 1971. Renamed Whitehorse Branch in 1977. Still active.
Yukon Branch	See Whitehorse Branch.
ALBERTA	
Calgary Branch	Established on November 10, 1917. First chairman: Lewis Stockett. Originally the Calgary Section of the Rocky Mountain Branch. Still active.

Name	Remarks
Drumheller Branch	Established with Council approval in March 1939. Drumheller was part of Rocky Mountain Branch in 1934. Discontinued in 1942.
Edmonton Branch	Originally established in 1912 as the Northern Alberta Group of the Rocky Mountain Branch. Became the Northern Alberta Branch in 1918 with L.E. Drummond (CIM Vice President in 1922-23) as first chairman. Suspended activities in 1956. Reactivated before 1967. Renamed Edmonton Branch with Council approval on March 20, 1981. Still active.
Grande Cache-Hinton	Established with Council approval on October 12, 1976. First chairman: Peter Clarke. Inactive.
McLeod-Athabaska Branch	Established at Edson on March 14, 1924. First chairman: J.H. McMillan. Ceased to exist after 1927.
Northern Alberta Branch	See Edmonton Branch.
Oil Sands Branch	Established at Fort McMurray in 1976. First chairman: A.C. Austin. Still active.
Rocky Mountain Branch	Established at Lethbridge on November 20, 1912. Originally encompassed Institute members in southeastern B.C., and southern Alberta as far east as Medicine Hat. Headquartered in Blairmore in 1923. Inactive from 1963 to 1973. Headquartered in Grande Cache in 1980. Still active.
NORTHWEST TERRITORIES	
Baffin Island Branch	Established at Nanisivik with Council approval on November 22, 1983. First chairman: W.M. John. Still active.
Pine Point Branch	Established on August 28, 1969. First chairman: H.P. Jacobi. Discontinued in 1981.
Port Radium Branch	Established in 1952. Discontinued in 1961.

Name	Remarks
Yellowknife Branch	Established with Council approval on October 5, 1945. First chairman: J.G. McNiven (CIM Vice President in 1959-60). Listed as active from 1967 to 1972. Reported inactive in 1973. Reactivated on October 15, 1974. Still active.
SASKATCHEWAN	
Beaverlodge Branch	Established at Uranium City in 1953. First chairman: E.N. Smith. Inactive from 1964 to 1968. Reactivated in 1968, and again in 1977. Discontinued in 1988.
Esterhazy Branch	Established on May 26, 1966. First chairman: J.B. Mitchell. Inactive from 1972 to 1978. Reactivated in 1979. Still active.
Regina Branch	See Saskatchewan Branch.
Saskatchewan Branch	Established as Regina-Estevan Branch in 1922. First chairman: A.A. Miller (CIM Vice President in 1924-25). Inactive from 1932 to 1937. Reactivated as South Saskatchewan Branch in 1938. Name changed to Saskatchewan Branch in 1952, Regina Branch in 1969, and Saskatchewan Branch again on September 6, 1986. Dissolved, as Regina Branch, by Council on October 17, 1988.
Saskatchewan Prairie Branch	Established at Estevan on November 15, 1965. First chairman: R.F. Palmer (CIM Vice President in 1966-68). There was an earlier Regina-Estevan Branch (see Saskatchewan Branch). Inactive since 1996.
Saskatoon Branch	Established in 1965. First chairman: R.F. Palmer. Still active.
South Saskatchewan Branch	See Saskatchewan Branch.

Name	Remarks
MANITOBA	
Flin Flon Branch	Established in 1933. First chairman: R.S. Jones (CIM Vice President in 1936-37). Still active.
Lynn Lake-Leaf Rapids Branch	Established as the Lynn Lake Branch with Council approval on March 22, 1971. First chairman: I.M. Plummer. Reorganized as Lynn Lake-Leaf Rapids Branch on May 4, 1973. Reactivated on October 8, 1983. Still active.
Manitoba Branch	See Winnipeg Branch.
North of 53 Branch	Established at The Pas on October 23, 1929. First chairman: Leon Dion. Disbanded in 1934.
Snow Lake Branch	Established on November 13, 1966. First chairman: J.R. Sadler. Still active.
Thompson Branch	Established in 1962. First chairman: I.P. Klassen (CIM Vice President in 1968-69). Still active.
Winnipeg Branch	Originally established 1918 as the Manitoba Branch at The Pas. First chairman: R.C. Wallace (CIM President in 1924-25). Name changed to Winnipeg Branch in February 1942. Still active.
ONTARIO	
Atikokan-Steep Rock Branch	Established on September 27, 1959. First chairman: K.K. Busby. Discontinued in 1964.
Algoma Branch	Established in 1957. Active in 1967. Reactivated in April 1980. Inactive since 1993.
Bancroft-Peterborough Branch	Established in 1959. First chairman: E.B. Wright. Also called Bancroft-Peterborough District Branch. Still active.

Name	Remarks
Central Patricia Branch	Established in 1956 as the Geraldton Branch. Renamed Geraldton-Central Patricia Branch in 1959. Chairman: D.S. Munro. Discontinued in 1973. Reestablished as Central Patricia Branch at Pickle Lake with Council approval on April 24, 1980. Inaugural meeting in 1981. Suspended operations in 1983.
Cobalt Branch	Established on April 15, 1907. First chairman: Arthur A. Cole (CIM President in 1916-18). Disbanded in 1930s. Reactivated in 1949. Still active.
Eastern Ontario Section	See Kingston Branch.
Elliot Lake Branch	Established at Elliot Lake on March 23, 1956 as the Algoma District Branch. Name changed to Elliot Lake Branch in 1959. Inactive. There was an earlier Blind River Branch.
Geraldton-Central Patricia Branch	See Central Patricia Branch.
Hamilton Branch	Established in 1953. Still active.
Hastings District Branch	Established at Deloro with Council approval on March 4, 1919. First chairman: S.B. Wright. Jurisdiction included Marmora, Madoc, Tweed, Sulphide, Kingston, Belleville, Fenton, Havelock and Cordova Mines. Discontinued by 1934.
Hemlo Branch	Established at Marathon with Council approval on October 4, 1984. First chairman: T.J. Stephens. Still active.
Ignace Branch	Established in 1977. First chairman: K. Blower. Dissolved by Council on September 30, 1991.

Name	Remarks
Kingston Branch	Established in 1902 as the Eastern Ontario Section. Renamed Kingston Branch in 1909. In 1919, Kingston was included in the Hastings District Branch. Kingston Branch reactivated in 1980, with inaugural meeting in 1981. Inactive since 1994.
Kirkland District Branch	See Kirkland Lake Branch.
Kirkland Lake Branch	Established on May 13, 1933. Meeting to discuss formation of a branch was held on October 12, 1932. First chairman: M.W. Hotchkin. Name changed to Kirkland District Branch on January 17, 1938, and back to Kirkland Lake Branch in 1992. Still active.
Lakehead Branch	See Thunder Bay Branch.
Lakehead-Geraldton Branch	See Thunder Bay Branch.
Manitouwadge Branch	Established on December 7, 1966. First chairman: L.S. Brooks. Inactive since 1996.
Michipicoten Branch	Established at Jamestown in 1954. First chairman: E.W.M. Cokayne. Disbanded in November 1969. Reactivated in 1981. Inactive since 1991.
Northern Gateway Branch	Established at North Bay in 1977. First chairman: A. Speirs. Inactive since 1996.
Ottawa Branch	Established on December 4, 1912. First chairman: W.R. Askwith. Still active.
Pickle Lake Branch	Established in 1944. First chairman: H. Brodie Hicks. Discontinued in 1952. Succeeded by Central Patricia Branch in 1980.
Porcupine Branch	Established on December 6, 1911. Still active.
Red Lake Branch	Established on August 25, 1943. First chairman: H.J. Bergmann. Still active.
Sudbury Branch	Established between October 6, 1911 and January 12, 1912. First chairman: R.G. Leckie (Institute Founder). Reactivated in 1919. Still active.

Name	Remarks
Thunder Bay Branch	Established at Port Arthur on November 11, 1935 as the Lakehead Branch. First chairman: H.A. Oakes. Later renamed Lakehead-Geraldton Branch, which was split in 1956 into Geraldton Branch and Lakehead Branch. Lakehead Branch renamed Thunder Bay Branch in 1970. Still active.
Toronto Branch	Established in 1906. First chairman: Eugene Coste (CIM President in 1903-05). Still active.
Werner Lake Branch	Established on March 14, 1967 at Kenora. First chairman: C.P. Moore. Discontinued in 1976.
	QUEBEC
Chibougamau Branch	See Chibougamau-Chapais Branch.
Chibougamau-Chapais Branch	Established in 1957 as the Chibougamau Branch. First chairman: D.M. MacLean (CIM Vice-President in 1959-60). Name changed to Chibougamau-Chapais Branch (Section Chibougamau-Chapais) in 1989. Still active.
Eastern Townships Branch	Established at Sherbrooke on June 10, 1902. First chairman: George R. Smith (Institute Founder, and President in 1905-07). Reorganized on August 27, 1910. Ceased to exist by 1923. A new Sherbrooke Branch was established on March 1, 1952, with J.O. Eby as chairman.
Gagnon Branch	Established in 1953. Operated as a section of the Northeastern Quebec Branch in 1963. Suspended activities in 1977. Dissolved in 1980.

Name	Remarks
Gaspé Branch	Established at Murdochville on March 18, 1963 as the Gaspé Peninsula Branch. First chairman: R.L. Coleman. Name changed to Gaspé Branch (Section Gaspé) in 1983. Still active.
Gaspé Peninsula Branch	See Gaspé Branch.
Harricana Branch	Established at Siscoe on March 4, 1936. First chairman: C.S. Johnston. Now called Section Harricana. Still active.
Joutel-Matagami Branch	Established on February 20, 1968. First chairman: D. Bridger. Now called Section Joutel-Matagami. Still active.
Knob Lake Branch	Established in 1963. First chairman: P. Munro. Dissolved on November 10, 1982.
Montreal Branch	Established on February 13, 1908. First chairman: George E. Drummond (Institute Founder). First vice chairman: John E. Hardman (CIM President in 1898-1900). Now called Section Montréal. Still active.
Northeastern Quebec Branch	See Québec Nord-Est Branch.
Québec Branch	Established at Quebec City on December 15, 1943 as the Quebec City Branch. First chairman: Théo C. Denis. Name changed to Québec Branch (Section Québec) in 1984. Still active.
Québec Nord-Est Branch	Established at Sept-Îles in October 1961 as the Northeastern Quebec Branch. First chairman: B.S. Imrie. Name changed to Québec Nord-Est Branch (Section Québec Nord-Est) in 1984. Still active.
Rouyn-Noranda Branch	Established at Noranda on June 29, 1929 as the Western Quebec Branch. First chairman: H.L. Roscoe (CIM Vice President in 1932-33). Name change to Rouyn-Noranda Branch (Section Rouyn-Noranda) approved by Council on March 20, 1989. Still active.

Name	Remarks
Saguenay Branch	Established at Chicoutimi with Council approval on May 8, 1975. First chairman: Jayanta Guha. Inactive since 1996.
Sherbrooke Branch	Established in 1952. The Eastern Townships Branch was based in Sherbrooke from 1902 to 1923. Ceased to exist in 1963.
Thetford District Branch	See Thetford Mines Branch.
Thetford Mines Branch	Established on December 4, 1920 as the Thetford District Branch. First chairman: George R. Smith (Institute Founder, and President in 1905-07). Name changed to Thetford Mines Branch in March 1930.Now called Section Thetford Mines. Still active.
Western Quebec Branch	See Rouyn-Noranda Branch.
NEW BRUNSWICK	
New Brunswick Branch	Original New Brunswick Branch established in September 1917. Present branch formed by amalgamation of Northern New Brunswick Branch and Southern New Brunswick Branch. Still active.
Northern New Brunswick Branch	Established in October 1955. First chairman: V.M. Glendenning. Amalgamated with Southern New Brunswick Branch.
Southern New Brunswick branch	Established at Fredericton on January 26, 1960. First chairman: C.S. Clements (CIM Vice President in 1965-67). Amalgamated with Northern New Brunswick Branch.
NOVA SCOTIA (MINING SOCIETY OF NOVA SCOTIA)	
Cape Breton Branch - Mining Society of Nova Scotia	Established on May 16, 1990. First chairman: Steve Farrell. Still active.
Halifax Branch - Mining Society of Nova Scotia	Established in 1936. First chairman: G.G. Bowser (CIM Vice President in 1944-45). Still active.

Name	Remarks
NEWFOUNDLAND	
Avalon Branch	Originally established as the Eastern or Avalon Section of the Newfoundland Branch in January 1976. First chairman: N. Kipnis. Avalon Branch approved by Council on March 18, 1983. Inactive since 1992.
Baie Verte Branch	Originally established as the Baie Verte Section of the Newfoundland Branch with Council approval on April 19, 1973. Inactive in 1981. Reactivated as Baie Verte Branch with Council approval on December 10, 1987. Still active.
Labrador Branch	Established in 1965 as the Wabush Lake Branch. First chairman: W.G. Taylor. Name changed to Labrador Branch in 1967. Also called Labrador West Branch. Council approved move from District 2 to District 1 on March 18, 1983. Still active.
Newfoundland Branch	Established at St. John's on November 25, 1953. First chairman: H.R. Peters. Still active.
Wabush Lake Branch	See Labrador Branch.
Western Newfoundland Branch	Established in December 1987 as West Coast of Newfoundland Branch. Renamed Western Newfoundland Branch in 1989. Inactive since 1993.

Notes: The available information on Branches is often incomplete. There is also some variation in reported dates.

CHAPTER THREE

TECHNICAL DIVISIONS AND SOCIETIES

The Institute's nine Divisions and Societies are the font of its scientific and technical expertise. Although specialized technical groups had existed since 1916, the first of the modern technical units began as the Industrial Minerals Section in 1932, and achieved Divisional status in 1941. More than ten years would pass before the next division, the Coal Division, was organized in 1944, quickly followed by the Metallurgy Division and the Geology Division in 1945. The Metal Mining Division was next, in 1946, and the Petroleum and Natural Gas Division was formed in 1949. These six Divisions appeared to serve the needs of CIM members adequately for the next twelve years. Then in 1961, members requested the formation of a Mechanical-Electrical Division. The Canadian Mineral Processors joined CIM in 1981, and the Mineral Economics Society was established in 1992.

Industrial Minerals Division

The Industrial Minerals Section came into being in 1932, after an incident at the 1929 Annual General Meeting in Winnipeg. At the meeting, a paper by L.H. Cole entitled "The Place of Non-Metallics in the Mining Industry" was never read as the session ran out of time.[454] When in 1931, this was brought to the attention of the Institute's new Secretary, E.J. Carlyle, he took steps to amend the oversight. A brief was prepared by interested members, and submitted to Council with a recommendation that a section devoted to furthering the non-metallic minerals industry be formed. The Industrial Minerals Section was sanctioned by Council on October 13, 1932.[455] L.H. Cole was appointed the first Chairman, and a Canada-wide committee was formed to organize the section. One of the original committee members was Frederick V. Seibert who would serve as Chairman of the Industrial Minerals Division in 1946-48, and as CIM President in 1949-50. The scope of the new Section was defined as being concerned with non-

[454] L.H. Cole and N.B. Davis, "History of the Industrial Minerals Division, C.I.M.," *Bulletin*, no 634 (February 1965): 217-218.

[455] CIM, *Council Minutes*, October 13, 1932, "The Industrial Minerals Section of the Institute," p. 143, and April 5, 1933, "Report of Committee, Industrial Minerals Section," p. 221, vol.4, NAC.

metallic minerals, clay products, and structural materials, but excluding fuels. At the Annual General Meeting (AGM) the following year, the Section hosted three sessions with 13 papers.

In the United States, the American Institute of Mining, Metallurgical, and Petroleum Engineers (AIME) followed CIM's example in 1935, and formed its own Industrial Minerals Division. Subsequently, the two groups held some joint meetings, and in 1965, a CIM member, M.F. Goudge (1951-52 CIM Vice President), was elected Chairman of the AIME group.

During World War II, the Industrial Minerals Section recommended to Council that the Institute press the Canadian government to establish "a Metals and Industrial Minerals Board . . . to mobilize the Canadian Mineral Industry [and] to assure adequate supplies of industrial minerals and metals required for the prosecution of the war."[456] This initiative eventually led to the government's appointment of George C. Bateman (1940-41 CIM President) as Canada's Metals Controller, and to the formation of the Institute's War Metals Advisory Committee.

In 1941, the Industrial Minerals Section asked Council for a name change to the Industrial Minerals Division (IMD). This was quickly approved by Council, "the service and growth of the Section having been so gratifying."[457] It thus became the first of the Institute's Technical Divisions.

Membership in the IMD in 1968-69 was 380, or 5.2 % of the Institute's total. In its annual report for that year, the Division stated that enrolment had been relatively constant for several years, despite the fact that large audiences had attended its technical sessions at the AGM. The Division's executive believed that there were a considerable number of people registered in other Divisions, who should also belong to the IMD.[458] This resulted in increased efforts to recruit new members, but participation in the Division remained fairly stable, with 330 members in 1996. In that year, after a period of relative inactivity, the IMD executive committee was reactivated with a mandate to expand CIM's presence in the industrial minerals sector.

Coal Division

Historically, interest in forming a coal section began after an informal survey revealed that 25% of CIM's membership identified with coal

[456] CIM, *Council Minutes*, "Communication from the Executive of the Industrial Minerals Section," April 13, 1940, p. 310, vol. 5, NAC.
[457] CIM, "Report of Council," *Bulletin*, no. 347 (March 1941): I-X.
[458] CIM, *Council Minutes*, "Industrial Minerals Division, 1968-69 Annual Report," 1969, vol. 8, NAC.

mining. A special committee was struck to study the matter, and a Coal Section, supported by the Institute and the Mining Society of Nova Scotia (MSNS), was in operation by 1919. The Section functioned like a branch, but there is little record of its activities. The idea of forming a Coal Division came from the MSNS in 1944. Council then invited the MSNS to discuss its proposal with coal men in western Canada. The MSNS sent its President, Alan E. Cameron, who was accompanied by the Institute's President, A.A. MacKay. The proposal was enthusiastically received, and a resolution for the formation of the Division was unanimously approved at the Annual Western Meeting in Vancouver. "The President, fully authorized by Council so to do, promptly declared the Division 'born' and designated Dr. Cameron its Chairman."[459] The Division was very popular from the outset, and Cameron was elected CIM President the following year (1945-46). Two other early Division chairmen would later also be CIM Presidents: Arthur E. Flynn (1953-1954), and David G. Burchell (1958-1959).[460]

In 1950, Burchell, R.D. Livingstone, and Alexis Ignatieff simplified the Division's structure. This allowed it to function more effectively, particularly with regards to stimulating national interest in coal-related problems. What the Division's organizers could not predict was the rapid deterioration of the Canadian coal industry, starting in 1953, which had a detrimental effect on the Division. According to its annual report for 1956-57:

> healthy Institute meetings by Divisions go hand in hand with the prosperity of the Industries represented. . . and likewise does the reverse create unhealthy Divisions. Coal, unfortunately has been in the latter category during the past few years.[461]

The Division worked hard to overcome these difficulties. Efforts were made to combine meetings, such as the joint technical session with the Metal Mining Division at the AGM in 1956, and the joint Solid Fuels Conference in 1957. Since 1959, the Coal Division had also established closer ties with its American counterpart, the Coal Division of AIME, by a cross-appointment of their chairmen as ex-officio executive members. The two divisions also organized joint technical sessions.

[459] "Establishment of the Coal Division," *Bulletin*, no. 392 (December 1944): 726-730.
[460] For a list of Division chairmen and secretaries from 1945 to 1984, see T.H. Patching, ed., *Coal in Canada*, CIM Special Volume 31 (Montreal: CIM, 1985) p. 327.
[461] CIM, *Council Minutes*, "Report of the Coal Division," April 25, 1957, minute 53, vol. 7, NAC.

The pattern of joint meetings continued with the Canadian Conference on Coal, co-organized by the Coal Operators' Association of Western Canada in 1970, and the first Open Pit Operators Conference, held in 1977, which brought together the Coal, Metal Mining, and the Mechanical-Electrical Divisions.

By 1976, the health of the Canadian coal industry had improved due to new international markets, but the Division's low membership, at only 4% of the Institute total, remained problematic. According to Walter Riva, Division Chairman in 1963-65, and CIM President in 1984-85, there was "strong pressure on coal to merge with one of the other Divisions." However, by the late 1970s, there was a marked increase in papers dealing with coal at the AGM, "and times when you couldn't get seats in the Coal Division meetings."[462] In 1981, the Division inaugurated its first Coal Operators Conference, which is normally held every two years, in a coal mining community.

In 1983, the Coal Division established the Coal Award for outstanding contributions to the industry and to the Division. The first recipient was Alexis Ignatieff, who had been the Division's Chairman in 1954-55, and had helped to keep the Division afloat through the hard times by his energy and enthusiasm.

Metallurgical Society of CIM

During Canada's Centennial year, the Metallurgy Division became the Metallurgical Society of CIM. Officially organized in 1945, its roots date to a much earlier period. In 1904, Eugene Coste (1903-05 CIM President) received several letters requesting the formation of a chemical and metallurgical section. He appointed a special committee to arrange the details, but a Metallurgical Section did not materialize until April 7, 1916.[463] During the same month, a second proposal was submitted for an iron and steel section. Council tried to interest the proponents in the existing Metallurgical Section, without success, and an Iron and Steel Section was approved on April 5, 1918.[464] Both it and the Metallurgical Section disappeared from the Institute's records well before World War II.

The recommendation for the formation of a Metallurgy Division came from a group of metallurgists in the Ottawa Branch. It was approved

---

[462] Walter Riva, interview with author, April 30, 1997, Parksville, B.C.

[463] CIM, *Council Minutes*, April 9, 1904, vol. 4, NAC. "Notes and Comments," *Bulletin*, no. 49 (May, 1916): 395.

[464] "Iron and Steel Section," *Transactions*, vol. XXII (1919): 8.

by Council on October 17, 1944, and a meeting to organize the Division was convened at the Annual General Meeting in Quebec City in 1945.[465] Its sponsors hoped to enhance:

> the influence of the Institute by providing a medium whereby metallurgical members could more readily exchange the results of their professional work and discuss problems arising therefrom.[466]

The first Chairman was Owen W. Ellis of the Ontario Research Foundation.[467] Reflecting the broad interest of its members, the Division organized four technical committees: Hydrometallurgy, Pyrometallurgy, Physical Metallurgy, and Electrometallurgy.[468] Among the Division's many distinguished members were Frank Forward who was Chairman in 1950-51, and CIM President in 1965-66, and Harold M. Wright, Chairman in 1952-53, CIM President in 1970-71, and a member of the Canadian Mining Hall of Fame.

From 1945 to 1967, interest in the Metallurgy Division strengthened. The Division was reorganized in 1961, and launched the first Conference of Metallurgists which was held in Hamilton in 1962. In support of this event, and to provide a venue for the publication of more metallurgical papers, the Mines Branch (now CANMET) initiated a new journal, the *Canadian Metallurgical Quarterly*. Although an official publication of the Institute, its production was underwritten by the Mines Branch, whose Director, John Convey, was the Conference Chairman.[469] Five years later, under the chairmanship of John H. Walsh, the Division submitted a request to Council that it be renamed the Metallurgical Society of CIM to reflect its new stature, and to compete on a more equal footing with the Metallurgical Society of AIME, its American counterpart.[470] This was approved on June 23, 1967.[471]

As the activities of the Society expanded, and the Conference of Metallurgists gained an international following, the Society was reorganized into seven technical sections in 1974. The number of sections

[465] CIM, *Council Minutes*, October 17, 1944, minute XVI, p. 52, vol. 5, NAC.
[466] "A Metallurgy Division," *Bulletin*, no. 394 (February 1945): 84-85.
[467] Fathi Habashi, "The Metallurgical Society Marks Fiftieth Anniversary," *Bulletin*, no. 999 (April 1996): 22-27.
[468] CIM, *Council Minutes*, October 5, 1945, minute 19, p. 142, and October 21-22, 1946, minute 25 (A) (4), p. 272, vol. 6, NAC.
[469] The Metallurgical Society subsequently assumed responsibility for the *Quarterly* and contracted its publication to Pergamon Press.
[470] John H. Walsh, interview with author, February 4, 1997, Ottawa.
[471] CIM, *Council Minutes*, June 23, 1967, minute 13, p. 1739, vol. 8, NAC.

has since grown to nine, the latest being the Management in Metallurgy Section formed in 1997. Since 1971, the Hydrometallurgy Section has held its own series of Annual Hydrometallurgical Meetings, either in conjunction with the Conference of Metallurgists, or separately. In addition to the technical sections, there are standing committees on the environment, computer applications, and historical metallurgy. The latter was formed in 1978-79 at the suggestion of Peter Tarassoff (1977-78 Metallurgical Society President, and 1990-91 CIM President), and has regularly published the popular "Historical Metallurgy Notes" in the *Bulletin.*

The core activity of the Society continues to be the Conference of Metallurgists which now features international symposia as well as workshops and short courses. The Society has also co-sponsored international meetings in other countries. A major strength of the Society is its publication program involving conference and symposium proceedings.

Over the years, the Society has established six awards: H.T. Airey (Noranda) Award (1963), Alcan Award (1966), Sherritt Hydrometallurgy Award (1976), Dofasco Award (1979), Silver Medal (1986), and Falconbridge Innovation Award (1989).

The Society currently has over 1,000 members, many of whom work in research and development related to the metals and materials industries.

Geological Society of CIM

The Geological Division began as a Geology Committee formed in 1942. Before then, Institute members had long participated in various external geological organizations. CIM Presidents Frank D. Adams and Alfred E. Barlow, among others, had been active in the Twelfth International Geological Congress held in Canada in 1913, and Section IV (Geology) of the Royal Society of Canada was chaired by Richard G. McConnell, an Institute Founder, in 1917-18.

A proposal for the formation of a Geology Division was received by Council in February 1942, and was approved in principle.[472] However, when the sponsors submitted the proposal to members attending the geological session at the AGM in March, it failed to gain majority support. Instead, those present recommended the formation of a Geology Committee, and it was duly approved by Council.[473] Its first Chairman was M.E. Wilson of the

[472] CIM, *Council Minutes*, February 7, 1942, p. 130, vol. 5, NAC.
[473] CIM, *Council Minutes*, March 12, 1942, p. 155, vol. 5, NAC.

Geological Survey of Canada, and the committee included such prominent geologists as A.H. Lang, A.E. Cameron, B.T. Denis, J.E. Gill, H.C. Gunning, W.J. James, P. Price and W.J. Wright. In 1944, the committee was referred to as the Mining Geology Committee, and its Chairman was another notable figure, Justin S. DeLury, who had been CIM President in 1930-31. About this time the idea of a Division was resurrected, and on October 5, 1945 the Mining Geology Committee was reconstituted as the Geology Division, with W.F. James as Chairman.[474] By 1946, it had evolved into a well organized and efficient Division attracting 995 members, under its Chairman James E. Thomson, Assistant Provincial Geologist for Ontario.[475] Eight committees and sub-committees had been created to deal with the various branches of geology, and a sub-committee was being organized to serve the interests of petroleum geologists. Work was also underway on The Structural Geology of Canadian Ore Deposits which was to be the first in the Institute's series of Special Volumes.

Progress throughout the 1950s and 1960s was marked by strong member participation. In 1956, the Division reported that nineteen papers were presented in its sessions at the Annual General Meeting. Three were given jointly with the Petroleum and Natural Gas Division, and six with the Geological Association of Canada (GAC). Some of the Division's most active members during these years included John F. Walker, Chairman in 1947 and CIM President in 1956-57, James B. Mawdsley, Chairman in 1948 and President in 1961-62, and Paul E. Auger, Chairman in 1960 and President in 1968-69.

By 1989, the Geology Division had 2,175 members, or 21% of the Institute total. The then Chairman, Chris Healey, acted on a recommendation by CIM President René Dufour that the Division prepare a position paper outlining its future objectives. Healey formed a special committee which recommended the organization of Annual Field Conferences, starting in the fall of 1991, and the publication of the Division's own geological journal.[476]

A sub-committee composed of Healey, Wayne Lockhart (Geology Division Chairman in 1989-90) , and John Jambor (Geological Society President in 1990-91), examined the journal's feasibility. During the deliberations, the question was raised as to whether the journal might have more credibility if it were published by a "Geological Society" instead of the Geology Divi-

[474] CIM, *Council Minutes*, October 5, 1945, minute 13(B), p. 137, vol. 5, NAC.
[475] "Progress of the Geology Division," *Bulletin*, no. 416 (December 1946): 697-98.
[476] John Jambor, "Notes on the Formation of the Geological Society," typescript, 2 p., 1996.

sion. Guided by Harlan Meade (Society President in 1991-92) , a change in name was accepted by the Division's members at the AGM in 1990, and was then approved by Council.[477] The change in status was timely, as it also supported the wider goals of the Division. The first issue of the Society's new quarterly journal, Exploration and Mining Geology, was published in January 1992 under a contractual agreement with Pergamon Press.[478]

The Geological Society administers the Barlow Memorial Medal (1926) awarded annually for the best geological paper published in CIM publications, and the Julian Boldy Memorial Award (1987) for exceptional service to the Society.

With two Sections, one in Saskatoon, and the other in Sudbury (1992), and a membership in the range of 2,000, the Geological Society is one of the largest and most successful of CIM's technical divisions.

Metal Mining Division

In November 1945, the Institute's Executive Committee convened a special committee to investigate the desirability of establishing a Metal Mining Division.[479] The committee was chaired by L.W. Bladon, Head of the Department of Mining and Metallurgical Engineering at McGill University, and its other members were J. Drybrough (1945-46 CIM Vice-President), A.E. Flynn (1953-54 CIM President), and R.V. Porritt (1944-45 Vice-President). Porritt would later become President of Noranda Mines, and a member of the Canadian Mining Hall of Fame. After consulting metal mining members of the Institute, the committee reported unanimous agreement on the formation of the Division but noted that "the general opinion was that it is a pity that divisions should have formed at all" and strongly recommended that "the breakdown of the Institute into technical divisions should not be carried into the Branches."[480] These sentiments were not surprising considering that mining members had always been the core of the Institute. The Division was approved by Council on April 12, 1946, and its first Chairman was H.L. Roscoe, Vice President of Noranda Mines, who had been a CIM Vice President in 1933-34. Bladon was "affectionately dubbed the 'Father' of the Division."[481]

---

[477] CIM, *Council Minutes*, May 10, 1990, minute 7(e).

[478] The Geological Society was guided by W.T. Thompson, then Chairman of the Central Publications Committee, who had been involved with the Metallurgical Society's *Canadian Metallurgical Quarterly*. The title of the journal was suggested by then CIM President Peter Tarassoff.

[479] CIM, *Council Minutes*, December 13, 1945, minute 43, p. 179, vol. 6, NAC.

[480] CIM, *Council Minutes*, April 12, 1946, minute 9(E), p. 245, vol. 6, NAC.

[481] "Officers of the New Metal Mining Division, C.I.M.M.," *Bulletin*, no. 412 (August 1946): 413-415.

Other Institute leaders who served as Metal Mining Division (MMD) chairmen included Ralph D. Parker, 1964-65 CIM President, Senior Vice President of Inco, and a member of the Canadian Mining Hall of Fame, and Richard E. Barrett, 1963-65 CIM Executive Director.

Much of the MMD's activities has focussed on the Annual General Meeting. At the AGM in 1961, the MMD organized a series of pre-session films on various aspects of the metal mining industry; this was so popular that the initiative was continued for several years. Starting in 1973, the MMD began organizing a mineral economics session at every second AGM, making it the first Division to embrace this field. It also organized joint meetings with other Divisions, and it was involved with the formation in 1974 of the Canadian Advisory Committee on Rock Mechanics, which was renamed the Rock Mechanics and Strata Control Committee in 1982.

The Division's first Underground Operators Conference was held in Sudbury in 1975, and due to the tremendous response, plans began almost immediately for the second conference in Val d'Or in 1976, which was co-hosted by the Harricana Branch. Participation was limited to 250 delegates, and like the first conference, the meeting was over-subscribed. This event has evolved into the biennial Mine Operators Conference. The Division also took part in the first Open Pit Operators Conference, held in the Labrador City-Wabush-Fermont region in 1977, with the Coal and Mechanical-Electrical Divisions.

The Metal Mining Division Award, inaugurated in 1983, recognizes outstanding achievements in the field of mining engineering.

Petroleum Society of CIM

The Institute and the industries it serves are inseparable, and the Petroleum Society's relation to the oil and gas industry is no exception. A Petroleum and Natural Gas Section had existed since November 1926, but interest tended to fluctuate with the fortunes of the industry.[482] Even though oil had been tapped in Turner Valley in 1914, and a larger reservoir had been found at Norman Wells in 1920, the oil boom in western Canada would not begin until the landmark discovery at Leduc in 1947. Like the other early technical sections of the Institute, the Petroleum and Natural Gas Section faded into obscurity.

[482] CIM, *Council Minutes*, March 22, 1926, p. 332, 339, vol. 4, NAC.

By the late 1940s, the Geology Division had been formed, and it had a Petroleum Geology Committee, Chaired by George Furnival. When Furnival was asked to survey Alberta geologists with regards to joining the Geology Division, he proposed that a new branch be established there. Instead, in 1949, Council approved the formation of a Petroleum and Natural Gas Division, with Furnival as its first Chairman.[483] Less than eight months after its establishment, the Division held its first Annual Technical Meeting in Edmonton.[484] The meeting included an excursion to Leduc and Woodbend, and did much to stimulate interest in the expanding oil and gas industry, as well as the new Division. The highly successful Annual Technical Meetings have since become the mainstay of the Petroleum Society's activities.

The Petroleum and Natural Gas Division's Chairman in 1950-51 was George Govier (1966-67 CIM President), who recalled that:

> after a few years, there was a friendly rivalry developing between the Society of Petroleum Engineers [SPE-AIME] in Calgary and the other technical people who were trying to expand the Petroleum and Natural Gas [Division]. . . In effect, the two groups were each striving to increase membership.[485]

Relations between the two groups had been fairly congenial through the 1950s. Council had approved a CIM-AIME Petroleum Engineering Section in Edmonton in 1957, and in the following year other sections were started in Calgary and Regina. Their purpose was to hold joint meetings. Then starting in 1960, more cooperation with the SPE-AIME was sought through Stan Harding, the Division's Chairman, and in 1966 a formal agreement was reached concerning membership recruitment. This arrangement ended in 1973 after the SPE decided to mount an ambitious membership campaign in Canada.[486]

In the interim, the Division had expanded its horizons. It supported the Calgary Technical Library opened in 1955, and held joint meetings with the Geological Association of Canada in 1954, and the Geol-

---

[483] George Furnival, telephone interview with author, November 1996, Calgary. R.W. Edgecombe, "The Petroleum Society of CIM — A Short History." *Directory,* (June 1981): 85-86.

[484] "C.I.M.M. Petroleum Division Meets in Edmonton June 1-3, Papers, Field Visits on Agenda," Oil in Canada, May 29, 1950.

[485] George Govier, interview with author, November 29, 1996, Calgary.

[486] CIM, *Council Minutes*, "Petroleum Society of CIM-SPE of AIME Agreement," April 19, 1973, minute 9, p. 2372.

ogy Division in 1956. Soon, the need arose for a special publication for papers dealing exclusively with petroleum and natural gas. This led to the launching in 1962 of the *Journal of Canadian Petroleum Technology (JCPT)* as a quarterly, and it rapidly became the principle technical publication for oil and gas in western Canada. These events, coupled with the need to compete more effectively with the SPE for new members, led to a change in stature for the Petroleum and Natural Gas Division. On April 1, 1965, Council approved a change in name to the Petroleum Society of CIM.[487] The new Society was ratified by the membership at the 16th Annual Technical Meeting in Calgary, with H.L. Bertrand as its first Chairman.[488]

The Petroleum Society has its own awards for Distinguished Service (1970), Outstanding Service (1985), the Best Paper published in the JCPT, and the Best Paper at the Society's Annual Technical Meeting. In 1986, the Society established its own administrative office in Calgary, and in 1993, publication of the *JCPT* was transferred from the Secretariat in Montreal.[489] Since its founding, the Society has probably contributed more candidates for CIM President than any other Institute Division or Society. With five Sections in Alberta, Saskatchewan, Ontario and Nova Scotia, and a membership of more than 2,000, the Society continues to grow and prosper as Canada's leading petroleum technology organization. Its contributions to the Institute's growth were formally recognized when the Institute changed its name to the Canadian Institute of Mining, Metallurgy and Petroleum in 1990.[490]

Maintenance/Engineering Division

In May 1954, CIM Secretary Jerry Gerow, received a persuasive letter proposing the formation of a new division:

> in addition to geologists, miners, and metallurgists, mechanical and civil engineers are also required for specific purposes in the mineral industry. We suggest consideration of a "Mechanical Civil Division." In each Division, there are a number of members whose duties require

[487] "The Annual General Meeting, C.I.M.," *Bulletin*, no. 636 (April 1965): 405.

[488] "The Petroleum Society of C.I.M.," *Bulletin*, no. 637 (May 1965): 569. "The Petroleum Society of C.I.M.," *Journal of Canadian Petroleum Technology*, vol. 4, no. 2 (April-May 1965): 4.

[489] Barbara Howes, "Everything You Wanted to Know About the Petroleum Society But Were Afraid to Ask," *Journal of Canadian Petroleum Technology*, vol. 33, no. 7 (September 1994): 11-12.

[490] B.J. McGuire, "Petroleum Comes in From the Cold," *Bulletin*, no. 940 (August 1990): 18.

> mechanical and civil qualifications and that a new Division, rather than being separate, would incorporate groups now active in the presently established technical Divisions.[491]

A 'Mechanical Civil Division' did not materialize, but in 1960 Council approved the establishment of a Mechanical-Electrical Division. It was organized on September 1, 1961, with a structure patterned after the Coal Division, and Donald J. McParland was elected its first Chairman, while L.O. Cooper was named Vice Chairman.[492]

During its first decade, the success of the M/E Division can almost be measured linearly. At its first AGM in 1961, nine technical papers were presented. In 1963, membership stood at more than 200. Two years later, it exceeded 300, and the Division promoted a new slogan called EMGAM, or the "Every member get a member". The Institute later adopted this slogan for a national membership drive. By 1967, the Division's membership topped 400. Eighteen technical papers were featured at the 1969 AGM. Starting in 1968, the Division co-organized a series of yearly presentations on major Canadian mining operations. The first was the Wesfrob Story, followed by the Wabush Story in 1969, the Griffith Mine Story in 1970, and the Brunswick Story in 1971. The Division's very successful biennial Mechanical-Electrical Operators Conference, now called Maintenance Engineering Conference, was begun in 1977.[493]

On October 4, 1984, Council approved a name change to the Maintenance/Engineering Division.[494] In his submission to Council, the Chairman of the Division, G.M. Behrend, noted that its interests and activities had become much wider than implied by the mechanical/electrical label. Currently, there are two M/E Sections in operation, one in Sudbury, and the other in Saskatoon.

Two awards are sponsored by the Division: the D.J. McParland Memorial Award (1972), and the J.D. (Pat) Patterson Memorial Scholarship (1991). The McParland Award honours the Division's first Chairman, who was killed in a plane crash while touring CIM Branches as a Distinguished Lecturer in 1969.

---

[491] CIM, *Council Minutes*, "Proposal to form a Mechanical Civil Division," November 25, 1954, vol. 7, NAC.

[492] Ernest Zucker, "History of the Mechanical-Electrical Division," *Bulletin*, no. 716 (December 1971): 23-24.

[493] Rick Clark, "The Maintenance Engineering Division of CIM," *Bulletin*, no. 1004 (October 1996): 13-17.

[494] CIM, *Council Minutes*, October 4, 1984, minute 27.

The Scholarship was established in memory of a long-time Division member who contributed greatly to its success. The Division's Distinguished Lecturer in 1978, Steven W. Harapiak, served as CIM President in 1985-86.

Canadian Mineral Processors Division

Among the Institute's nine Divisions and Societies, the Canadian Mineral Processors Division is unique in having been a full-fledged, independent organization before it joined CIM in 1981. Its origin lay in a meeting of gold mill superintendents organized at Porcupine in 1957 by Jack Carter of McIntyre Porcupine Mines, Ian Gordon of Hollinger Consolidated Gold Mines, and Russel Sullivan of Preston East Dome Mines.[495] The purpose of the meeting was to discuss technical problems of mutual interest in an informal atmosphere, and attendance was limited to two representatives from each company. Its success prompted the formation of similar groups in the Kirkland-Larder Lake, Malartic-Val d'Or and Red Lake areas.

The Mines Branch saw these meetings as an opportunity for orienting its research programs to the benefit of the gold industry, and Ken Downes, the Chief of the Extraction Metallurgy Division invited mill superintendents from all Canadian gold mines to a meeting in Ottawa in January 1962. Its success prompted the attendees to request another meeting the following year, which led to the formation of the Canadian Gold Metallurgists. By 1967, its Annual Operators Conference had become well-established, with the support of the Mines Branch, and mill operators from base metal mines were invited to participate. The group was then renamed the Canadian Mineral Processors (CMP), and in 1970, representatives from the iron ore industry were also included.

During 1970, the CMP discussed possible affiliation with CIM, but the Institute's proposal was conditional on CMP becoming part of the Metallurgical Society. This idea was rejected because the CMP members wanted to maintain their focus on operating issues, as well as retain their autonomy.[496] Ten years later, Don Schmitt, the 1980-81 CIM President, accompanied by George Smith, President of the Metallurgical Society, and Gordon Skilling, CIM Executive Director, met with a CMP delegation led by its Chairman Phil Claridge.[497] Terms were agreed upon for the CMP to

---

[495] R.J.C. MacDonald, "The Canadian Mineral Processors - A Brief History," typescript, 33 p, May 1997.

[496] Richard Fish, "Mineral Processors Reject CIM Affiliation this Year," *Canadian Mining Journal*, vol. 100, no. 3 (March 1979): 35.

[497] CIM, *Council Minutes*, September 17, 1980, minute 24.

join CIM as a Division, and were ratified by the CMP membership in January 1981. The new Canadian Mineral Processors Division was approved by Council on March 20, 1981.[498]

Membership in the CMP has increased from 50 in 1964, to 750 in 1997. The Division now brings together mill operators from all sectors of the mining industry, as well as researchers, consultants, academics and suppliers. The CMP maintains ties with the Metallurgical Society's Mineral Sciences and Engineering Section, and has co-sponsored technical sessions on mineral processing at the Conference of Metallurgists and at the Annual General Meeting.

CMP has initiated two major annual awards: the Mineral Processor of the Year Award (formerly the Mill Man of the Year Award, established in 1984), and the Bill Moore Special Achievement Award (formerly the Special Achievement Award, 1984). It also sponsors an award for Best Presentation at the AGM (1990).

Mineral Economics Society of CIM

The Mineral Economics Society began as a special committee on mineral economics. In 1973, Council discussed a proposal for the formation of a "Council of Economics similar to the one operated by AIME and to which CIM names a Canadian representative each year."[499] Although the proposal was considered to have merit, it was questioned whether the Institute had the resources for such a council, and it was also noted that the Metal Mining Division already had a session on economics at alternate Annual General Meetings. The following year, Robert B. Elver asked Council to again consider the establishment of a mineral economics committee or division. In a thoughtful and rather prophetic position paper on "The Role of Mineral Economics in the C.I.M." it was observed that:

> Never before have the issues of mineral policy and the many questions pertaining to how mineral resources should be privately or publicly managed been the subject of such wide-spread political debate both within and outside Canada"[500]

[498] CIM, *Council Minutes*, March 20, 1981, minute 33.
[499] CIM, *Council Minutes*, April 19, 1973, "Mineral Economics Committee Proposal," minute 52, p. 2220.
[500] CIM, *Council Minutes*, April 25, 1974, "Mineral Economics Committee Proposal," minute 6, p. 2085.

The paper went on to say:

> CIM's membership represents a national asset that could make a more substantial contribution to the full range of on-going public debate bearing upon the use and development of Canada's mineral resources.....there is an opportunity for CIM to enhance its role.....and not only could the CIM make a more authoritative contribution to national mineral policy development, but it could also incur professional enrichment and satisfaction for many of its members.

Council heartily approved the proposal and authorized the formation of a Mineral Economics Committee. It consisted of representatives from each technical Division and Society, and from the Institute's membership at large, and its first Chairman was Robert Elver. The Committee's goals were to act for the Institute in the development and transfer of information on mineral economics, and to encourage an interdisciplinary approach to broad mineral issues among Institute members.

In 1977, the Committee reported that in its third year of operation, it had organized two sessions on mineral economics at the Annual General Meeting, and had assisted in a third session sponsored by the Geology Division. These had been well attended. Donald J. Worth (1996-97 CIM President), who presented the Committee's report to Council, also said that the Committee's plans for 1978 included a joint session with the Mechanical/Electrical Division on equipment economics.[501]

In 1980, the Committee organized its first Mineral Economics Symposium. With the theme "Perspectives on Mineral Evaluation and Forecasting," the two-day meeting attracted 185 registrants. The Symposium, held every two or three years, quickly gained an international following. It was normally held in Toronto, a major corporate mining industry centre, but in 1992, back-to-back meetings were organized in Toronto and Vancouver. This innovative approach to reaching CIM members across Canada has since been repeated.

Building on its success, in 1992 the Mineral Economics Committee petitioned Council for divisional status. The proposal was presented by the Committee's Chairman, Al Clark, and John Lydall, a Past Chairman, and approved by Council after agreement that the originally proposed

[501] CIM, *Council Minutes*, "Mineral Economics Committee," April 21, 1977, p. 2568.

name "Mineral Economics and Management Society" be shortened to the Mineral Economics Society of CIM.[502]

The memory of Robert Elver, who was instrumental in launching the Mineral Economics Committee in 1974, is commemorated by the Robert Elver Mineral Economics Award established in 1981.

***

As 1998 unfolds, the technical Societies and Divisions continue to focus on CIM's Strategic Goal of the transfer of knowledge.

[502] CIM, *Council Minutes*, January 20, 1992, "Proposal from Mineral Economics Committee," minute 6.

CHAPTER FOUR

THE ANNUAL GENERAL MEETING

The rich tablecloths offset the burgundy wine glasses, white napkins, and silver cutlery of each place setting. Suspended over the head table, the Institute's royal-blue and gold crest was accompanied by historic photographs, each portraying a person or an event from Canada's minerals industry heritage. As four Scottish pipers led CIM's officers into the banquet hall, followed by flag bearers carrying the Canadian, British Columbia, and Quebec colours, hundreds of delegates and their guests rose in respect. After the head table party had taken their seats, everyone was invited to join in a toast 'to Canada.' Thus, the Business Luncheon of the 99th Annual General Meeting was officially opened, and participants waited for the Presidential address, before the serving of an elegant meal. The proceedings could have been from almost any other Annual General Meeting, were it not for the absence of a toast to the Monarchy, the modern style of dress, and the timely topics of conversation. This was Vancouver, 1997, and the last official gathering before the Institute's Centennial.

Traditionally, the Annual General Meeting has always been the highlight of the Institute's year. To quote David Robertson, President in 1993-94:

The AGM is something of a microcosm of CIM:

> it provides fellowship, technical exchange and a learning experience for all. Much like the Institute itself, it seeks to bring together people from across the country and around the world to learn about Canada's mineral industry.[503]

Because the Technical Divisions and Societies showcase the latest in their fields of endeavour, the AGM technical sessions continue to be a means for the interdisciplinary exchange of knowledge. The trade exhibition, which harks back to 1926, provides CIM members and industry suppliers an opportunity to mingle and discuss the latest in technology.[504] In recent years, Tradex, with its elaborate displays and large pieces of equipment, has become a permanent feature of the AGM, and its policy of inviting the public to a one-day open house

[503] David Robertson, "Observations by the President," *Bulletin*, no. 978 (March 1994): 8.

[504] "Exhibit of Minerals, Mineral Products, and Mining Equipment," *Bulletin*, no. 169 (May 1926): 560.

has promoted a wider appreciation of Canada's minerals industry. The added attraction of a large city, and a variety of social events at the AGM, provides a welcome change of pace for CIM members and their guests. At the registration desk, or under the door of hotel rooms in the morning, the *CIM Reporter*, introduced in 1975, highlights the events of the day in English and in French.

Although many aspects of the AGM are throwbacks to the past, some of its original flavour has receded with the years. During the Institute's first decades, members who could not attend the AGM could read a detailed account in the *Bulletin*. It was the next best thing to actually being there. Speeches and discussions were often printed verbatim, and field trips and other informal activities were recounted in vivid detail. At one of the infamous 'smoking concerts', chaired by Colonel J.J. Penhale in 1922, several "husky song leaders divided the audience into two opposing choirs and managed to conduct both choral and duet competitions, which probably impressed the ordinary garden variety of hotel guest..." The Secretary-Treasurer, George C. Mackenzie, sang "the whole 'steen verses of the Institute Battle Hymn" (Drill, Ye Tarriers, Drill), followed by "a clever series of lantern slides, depicting various Institute Notabilities in startlingly life-like attitudes..." These were accompanied by limericks describing each person's professional alliance. At the conclusion of the entertainment, members adjourned for refreshments in order to fortify themselves "for the occasional and unofficial relays of smoking concerts, which continued until daylight..." and for several days afterwards.[505]

Good fun, great food, and fellowship have always been synonymous with the AGM, although in more recent times, delegates have tended to curtail their boisterous enthusiasm. Some can still recall the cribbage games and the sing-alongs in the hotel hospitality suites. In earlier years, the songs would come from the Institute's 'official' Hymn Book, Hymns, Ancient and Modern, but Mostly Archean: the Only Authorized Version, which contained rollicking verses like those found in the "Cobalt Song":

> For we'll sing a song of Cobalt, If you don't live there it's your fault,
> Oh you Cobalt where the wintry breezes blow, Where all the silver comes from
> And you live a life and then some, Oh you Cobalt, you're best town I know.[506]

[505] "The Smoking Concert," *Bulletin*, no. 120 (April 1922): 462-63.

[506] L.F. Steenman, words, R.L. MacAdam, music, "The Cobalt Song, published by Whaley," Royce & Co. Ltd., 1928.

Elaborate cuisine and extraordinary dinner menus often made for excellent dining and stimulating conversation at the Institute's Annual Dinners. In 1911, the Asbestos Manufacturing Company supplied a menu printed on an asbestos shingle, and in 1932, the menu was on a sheet of pure copper. At the 1936 AGM, the menu enclosed a poem and artwork by S. C. Ells, along with the complete verses of "The Miner's Song" by Norman B. Davis. Needless to say, delegates and guests were expected to sing before their entrées. Very occasionally, a meal would be less than memorable. At the 1966 AGM held at the Chateau Frontenac in Quebec City, members were served a luncheon of partially boiled potatoes, cold canned corn, and cold ham after the kitchen's boiler had exploded.

At times, the annual meetings took place under more solemn circumstances. In 1904, social events and field trips were cancelled after news was received of the death of B.T.A. Bell, the Institute's esteemed Secretary. Just after World War I, in place of the customary smoker, the evening's entertainment consisted of film, never previously shown, of the Canadian armies on the Western Front. And in 1919, members of the Montreal Branch suggested that in view of a small pox quarantine in Ottawa, the 21st AGM should be moved to the Windsor Hotel in Montreal, if other eastern branches were in agreement.[507] This was done. Worldly concerns did not dampen the proceedings at the 39th Annual General Meeting in Toronto in 1938. Despite the distractions of "Herr Hitler and the stock market," the meeting was reported to be the best attended and "jolliest convention in the history of the Institute."[508]

Formalities at the AGM would also fluctuate. In 1936, Captain James Ross suggested that it was time to abandon the "antiquated institution of an elaborate head table" at luncheons and dinners. But the tradition of a the head table endured, and in 1965 Council officially approved a policy of "black tie" as the accepted dress for the head table at the Annual Dinner. This added an air of solemnity to the presentation of the Institute's prestigious honours and awards.

Although times have changed, and Institute meetings have taken on a more 'professional' atmosphere, the AGM can still have its exciting moments. Noel Cleland recalled that at the 1981 AGM in Calgary when he was President-Elect, he and Donald Schmitt, the President, R. Heath Gray, the President-Elect designate, and Gordon Skilling, the Executive Direc-

[507] "Montreal Branch," *Bulletin*, no. 107 (March 1921): 241-2.
[508] "Thirty-Ninth Annual General Meeting," *Bulletin*, no. 312 (April 1938): 161.

tor, and their wives, were on the 29th floor of the hotel when the fire bell rang. After a few tense moments, and the appearance of several fire trucks, the verdict was a false alarm.[509] However, the incident left them a little shaken. Apparently, nobody has ever been lost at an AGM, but one delegate did get rather close. At a meeting in Vancouver, a member who was known for his over-indulgence disappeared one evening to the distress of his colleagues. After several hours of searching, an anxious call was made to the police, who almost dredged the harbour. The member finally reappeared in the morning, with a hangover, but body and soul intact.

As the 1997 Business Luncheon wound to a close, the President, Don Worth, handed the responsibility for the Institute's leadership to A.M. (Sandy) Laird, and was in turn presented with the traditional gift for the retiring president. Almost one hundred years ago, in 1898, George Drummond, Chairman of the Federated Canadian Mining Institute had similarly passed the torch to John E. Hardman, President of the newly organized Canadian Mining Institute. Laird's theme for 1997-98 was relevance, and how it applies to the membership:

> Our engineers no longer live in isolated company towns … and they have different lifestyles and priorities than they did in the past… We must focus on serving the interests of these younger members by offering them highly focussed and exciting programs and greater opportunities to exchange knowledge, to meet their peers, and to engage in good fellowship.[510]

This general subject has been one of the recurrent threads throughout the Institute's history. How can the Institute continue to grow and remain relevant to its members and diverse constituencies? Part of the answer can be found in a vibrant Annual General Meeting that showcases the latest achievements in the 'arts and sciences' of the minerals industry, and challenges the industry's professionals, often with surprising and lasting results.

---

[509] Noel Cleland, interview with author, January 14, 1997, Calgary.

[510] Sandy Laird, "CIM Strategic Plan Demands Relevance and Commitment," *CIM Reporter*, April 29, 1997, p. 1.

EPILOGUE

The Institute's Centennial celebrations officially began on January 1, 1998, although the main feature will be the 100th Annual General Meeting in Montreal during the first week of May. The joint participation of the Council of Mining and Metallurgical Institutions (CMMI), a group of twelve sister societies from around the world, and the Multilateral Investment Guarantee Agency (MIGA) of the World Bank, a corporation which promotes private direct investment in emerging countries, will make the celebration a truly international event. With more than 400 technical papers scheduled for presentation, eight special seminars and symposia on current topics, and a Trade Show with over 350 manufacturers and suppliers, CIM/CMMI/MIGA Montréal '98 is destined to be a meeting of a magnitude unprecedented in the Institute's history.

A schedule of special events is currently unfolding across Canada. But perhaps twenty-five years from now, the person who will review this CIM landmark will speculate on the highlights, and the tremendous amount of personal effort required to make it so successful. When all the statistics have been compiled, and the proceedings duly recorded, will there be enough anecdotal material to give an accurate picture of the times? In this age of electronic communication, will anyone actually write about the Centennial, and express his or her feelings, along with a few interesting stories about people and happenings? The past can only be made tangible to the next generation if enough sources have survived to make interpretation possible. In its 100th year, the progress of the Canadian Institute of Mining, Metallurgy and Petroleum certainly deserves the best attention that its members can bestow.

APPENDIX I

INSTITUTE FOUNDERS AND PRESIDENTS

The Founders

Benjamin Taylor A. Bell

See Institute Secretaries, Appendix II.

William Blakemore

D. 1917. Employed by Crow's Nest Coal in (Fort) Mcleod, BC, in 1898. One of the first Institute members to report on the Frank Slide disaster in 1903. Awarded a prize for a valuable paper submitted to the Proceedings of the North of England Institute of Mining and Mechanical Engineers on "Coal Cutting by Machinery." **CIM:** Councillor in 1898; active in Institute affairs.

Archibald Blue

B. 1840, Oxford, Kent County, ON. d. July 27, 1914, Ottawa, ON. Obtained a high school diploma, was a school teacher, and then a journalist. Wrote reports and articles for the St. Thomas Journal, and in Toronto, the Globe, and the World. Joined the Ontario Government as a civil servant in 1882. Organized and became secretary of the Bureau of Industries (the first provincial body to gather statistics on agriculture). In 1884, was appointed deputy minister of agriculture. In 1891, became the first director of the Bureau of Mines, established the same year. Resigned this post in 1900 to become chief census commissioner of Canada. A great believer in the value of gathering all kinds of information. He had an encyclopedic memory of the mining industry. **CIM:** Councillor 1898-99.

John Blue

D. 1907. Managed the Eustis Mining Co., a copper producer in Capleton, QC. CIM: President of the General Mining Association of the Province of Quebec 1895. **CIM:** Vice President 1898-99; Councillor 1905-06.

William A. Carlyle

B. 1862, Hamilton, ON. d. 1947. Graduated from McGill University in 1887 with BASc in mining and metallurgy. Spent four years in silver mines near Aspen, CO, returned to McGill as a lecturer. Became British Columbia's first provincial mineralogist and director in 1895. Manager of mines at Rossland for the British American Mining Corp. for two years. Left Rossland in 1900 to be general manager of the copper mines at Rio Tinto, Spain. From 1909 to 1913, he was professor of metallurgy at the Royal School of Mines, London. In 1913, retired as a consulting engineer and travelled extensively. In 1916, he returned to Canada and was managing director of the British America Nickel Corp. at Sudbury. Equipped and reopened the Murray mine, built a smelter at Nickelton, and a refinery at Deschenes, QC. **CIM:** Vice President 1898-99; Chairman of the Ottawa Branch 1918;. Councillor 1919. He had resigned as a Charter Member, but rejoined the Institute on his return to Canada in 1916. Member of IMM, and the Institution of Civil Engineers. An elder brother to E.J. Carlyle, who was Secretary-Treasurer of CIM from 1931 to 1949.

George Mercier Dawson

B. August 1, 1849, Pictou, NS. d. February 28, 1901, Montreal, QC. Entered McGill University in 1868 but transferred to the Royal School of Mines in London. Graduated in 1872 with distinction in geology and mining. Joined the Geological Survey of Canada in 1875, was director from 1895 to 1901. Received a DSc from Princeton University in 1877, LLD from Queen's University in 1890 and the University of Toronto in 1899. Son of Sir William Dawson, geologist and principal of McGill University. Had a wide knowledge of natural history. Keenly interested in the languages and cultures of North American aboriginal people. Dawson City was named in his honour. Awarded the Bigsby Gold Medal of the Geological Society in 1891. Fellow of the Royal Society of England in 1891. Elected president of the Royal Society of Canada in 1893, and president of the geological section of the British Society for the Advancement of Science. President of the Geological Society of America in 1896. "Canada's foremost scientific mind of the latter nineteenth century." **CIM:** Vice President 1898-99; active in Institute affairs.

Clarence Hoyt Dimock

B. October 25, 1855, Windsor, NS. d. February 15, 1930, Windsor, NS. Operated gypsum quarries at Windsor with his brother. They shipped gypsum to the J.B. King Co. of Staten Island, NY. Clarence became president of the Wentworth Gypsum Co. of Windsor. Director of the Commercial Bank of Windsor, and of the Windsor Foundry Co. Interested in gold mining. Widely known and admired. An impeccable dresser, and an all-round athlete. Once very wealthy, but succumbed to ill health and financial misfortune in later years. **CIM:** Councillor 1898-99.

George E. Drummond

B. 1829. d. 1919. Manager of the Canada Iron Furnace Co. in Montreal, QC. Became a Senator. **CIM:** Councillor 1898-99; Vice President 1899-1900, 1908-09; first Chairman of the Montreal Branch 1908.

Charles Fergie

See Presidents.

John E. Hardman

See Presidents.

John B. Hobson

B. 1844, Ireland. d. January 9, 1912, Victoria, BC. Arrived in New York with his parents in 1848, moved to California in 1857. Studied mining engineering and metallurgy. One of several men who formed the California Miners Association. Associated for many years with the Geological Department of the California State Mining Bureau. In 1892, was invited by Sir William Van Horne and the Canadian Pacific Railway to visit northern British Columbia to explore and assess the region. In private practice in Quesnelle Forks, BC, with L.F. Warner, in the Cariboo District, citing twenty years of practical experience in California. Specialized in deep gravel, drift, hydraulic, and gold quartz mines. A pioneer in the field of hydraulic mining, he invented a "deflecting nozzle" which became known as "Hobson's Improvement." **CIM:** Councillor 1898-99, 1901-02; Vice President 1903-04. Member of AIME.

Robert Gilmour Leckie

B. August 23, 1833, Renfrewshire, Scotland. d. November 5, 1913, Sudbury, ON. Educated at Glasgow High School and Glasgow Technical College. Came to Montreal at age twenty-three, and worked in shipbuilding. Then became interested in mining, first in Quebec's Eastern Townships, where he was associated with the Orford Nickel and Copper Co., then in gold and coal in Nova Scotia. About 1880, he was partly responsible for the consolidation of coal companies operating in the Springhill area, in association with the Springhill and Parrsborough Railway Co., which emerged as the Cumberland Railway and Coal Co., of which he was managing director. His managerial abilities were so consistent that during his appointment, the company was free from strikes. In 1890, he became general manager of the Londonderry Iron Co., and helped consolidate other coal mines in the region. Left Nova Scotia in 1898, becoming a mining consultant for an American firm on nickel and other minerals in New Caledonia, Norway, Sweden, and Australia. Was then resident consulting engineer for the Canadian Copper Co. in Sudbury. After retirement, he acquired and partially developed mining prospects in northern Ontario, including a gold property at Long Lake, near Sudbury. Sometimes regarded as "the father of mining engineering in Canada." **CIM:** Helped to organize the MSNS; first President of the Federated Canadian Mining Institute; CIM Councillor 1898-99, 1901, 1904; Vice President 1905-6. Member of AIME, and Vice President in 1893-94. At the annual general meetings, he was the "life and soul" of the gathering, and despite his age, was considered by many as the "youngest man" there.

Wilbur L. Libbey

B. circa 1865, United States. d. 1925, Boston, MA. In 1894, was president of the Brookfield Mining Co., North Brookfield, Queens County, NS. A gold-bearing fissure-vein bears his name. At the time, he sank the deepest vertical shaft in the province (1,062 feet). Introduced both the chlorination (1897) and bromo-cyanide (1903) processes for gold extraction while employed by H.S. Badger. When the mine closed in 1905, because of technical problems and loss of capital, he worked in Mexico in coal mining. **CIM:** President MSNS 1900-02; CIM Councillor 1898-99, 1902. "A big, broad-shouldered, black-haired, handsome man who always presided at [the MSNS] meetings in a faultless frock-coat."

Frank C. Loring

B. 1860, Valparaiso, Chile. d. November 2, 1939, Toronto, ON. Moved to Colorado at an early age with his parents, who were originally from New England. A mining engineer at Josie Gold Mining Co. in Rossland in 1898. Among the earliest to work in the Cobalt district in 1904, and about 1909, in Kirkland Lake. Had a pioneering spirit. **CIM:** Councillor 1898-99; Life Member.

Albert Peter Low

D. October 9, 1942. Graduated from McGill University. Joined Geological Survey of Canada in 1882, appointed director in 1906. Became first deputy minister of the federal Department of Mines in 1907-14. Retired from public life in 1942. Best known for his scientific expeditions to the Labrador Peninsula, Hudson's Bay and Arctic Islands. Especially good at organization, but it was his field exploits that won him admiration. In his obituary, it was written that he was "quick and strong, of immense energy and perserverence, with a constitution that could withstand all hardships, coupled with an intense interest in all branches of natural history. He had the consuming curiosity of the born explorer which must always see the other side of the hill or the other end of the river."

James McArthur

B. Scotland. A furnace man trained by Sir Henry Bessemer, he immigrated to Canada. Employed by the Orford Nickel and Copper Co. at the Orford nickel mine in Quebec's Eastern Townships, circa 1878. Later joined the Canadian Copper Co. in Sudbury, ON. **CIM:** Councillor 1898-99; Vice President 1900-01.

Richard George McConnell

B. March 26, 1857, Chatham, QC. d. 1942. Studied at McGill University obtaining a BA in natural science in 1879. Appointed to the Geological Survey of Canada in 1892 as an assistant to George Dawson. Explored and mapped vast regions from British Columbia to Manitoba, northern Quebec, and the Yukon, particularly the hazardous Stikine, Mackenzie, Porcupine, and Liard rivers. In 1898, succeeded W.A. Carlyle as provincial mineralogist for British Columbia. Appointed federal deputy minister of mines in 1914 until retirement in 1920. A leading authority on the geology and mineral resources of western Canada. His most notable scientific work was in connection with igneous and metamorphic rocks, and ore deposits, in BC and the Yukon. His only son became a notable geologist. **CIM:** Councillor 1900-01; Honourary Life Member. In 1917-18, he was President of Section IV (Geology) of the Royal Society of Canada. Fellow of the Royal Society in 1913. Frank Dawson Adams described McConnell as "a strong, rugged, and intrepid explorer, robust and of sturdy Scots ancestry, great reader of good books, and a favourite with all whom he associated."

William Hamilton Merritt

B. 1855, St. Catharines, ON. d. October 26, 1918, Toronto, ON. Educated at the Royal School of Mines, London, graduated in 1877. Canada's official representative to the Paris Exposition of 1878. Appointed lecturer in mining engineering at School of Mines in Kingston in 1890, served on Ontario Royal Commission investigating Ontario's mineral resources. In 1898, was in private practice as a mining engineer and metallurgist. Spent three years in active service in South Africa during the Boer War. In 1903-08, commanded the Governor-General's Bodyguard. In 1911, appointed to command the 1st Cavalry Brigade. Took an active part in the training of aviators in World War One. Attained rank of Lieutenant-Colonel. Author of numerous papers. Died during the influenza epidemic of 1918. **CIM:** Councillor 1898-99.

Joseph Obalski

B. 1852, France. d. March 25, 1915, Montreal, QC. Educated at l'École des Mines in Paris. Owned a consulting practice in mining engineering in Montreal from 1877 to 1915. Was Quebec's mine inspector from 1881 until his retirement in 1909. Interested in education. Elected president of the Chambre de Commerce Française in 1914-15. **CIM:** Councillor 1898 and 1905; Vice President 1898, and 1909-10.

John Jenkins Penhale

B. 1866, Leedstown, Cornwall, England. d. August 28, 1926, Thetford Mines, QC. Educated at the Centenary Collegiate Institute at Hackettstown, NJ.

Both he and his father, Mathew Penhale, a Cornish mine captain, were closely associated with the Quebec asbestos industry, and they designed and built the world's first asbestos mill at Black Lake in 1887-88. John Penhale also designed an experimental mill for the United Asbestos Co. in 1890. In World War I, was Lieutenant-Colonel in the First Division and took part in operations at the Front. After Armistice, went to Germany with the army of occupation. Mentioned in dispatches in 1916 and 1918, awarded the DSO. After returning to Canada, was postmaster at Sherbrooke until 1924, then assistant manager of the Asbestos Corp. **CIM:** Played an active role in the formation of the General Mining Association of the Province of Quebec in 1891; CIM Councillor 1898-99; Chairman of the Flanders Branch, in Belgium; except for the war, was a Councillor from 1910 to 1926; Life Member in 1924. "He occupied an enviable position in the esteem and affectation of members from coast to coast." At social gatherings, the "smoker was not considered complete without the genial Colonel in the Chair."

Henry Skeffington Poole

B. 1844. d. 1917, England. Educated at King's College, Windsor, then studied at the Royal School of Mines in London. Returned to work under his father at the Caledonia Colliery in Nova Scotia. In 1872, Poole became the manager of Acadia Colliery at Westville, which amalgamated into one company encompassing the Vale and Halifax coal mines. Poole remained manager. One of the most experienced coal mining men in eastern Canada. **CIM:** President of the MSNS 1892-94; CIM Councillor 1900-01.

George R. Smith

See Presidents.

A.W. Stevenson

Chartered Accountant, Bank of Toronto Chambers, Montreal **CIM:** Treasurer of the General Mining Association of the Province of Quebec; CIM Treasurer 1898-99.

Frederick T. Snyder

Employed by the Ottawa Gold Milling and Mining Co., Keewatin, ON. Probably moved to Chicago sometime before 1908 to work in the steel industry. Designed an electric furnace for use in steel manufacturing, a modified version of one invented by Girod, it was aptly called the "Snyder Electric Furnace." **CIM:** Councillor 1898-99.

The Presidents

John E. Hardman, 1898-1900

B. 1856, Lowell, MA. d. April 3, 1925, Montreal, QC. Studied at Massachusetts Institute of Technology. Obtained BSc in mining engineering. Postgraduate studies at the Royal School of Mines, London; l'École des Mines, Paris, Berg Akademie, Berlin. Worked as a chemist and metallurgist with the American Smelting Co. in Colorado. Superintendent of Hibbard Antimony Co. Mine owner and operator, 1881-95. Arrived in Nova Scotia in 1883. Worked in the Oldham and West Waverly gold districts near Halifax, 1884-94. Lectured at McGill University for one term. Consultant in mining engineering from 1895, specializing in gold mining and milling. A polished and persuasive speaker. **CIM:** Charter Member of the Gold Miners Club; President MSNS 1894-95; Institute Founder; President 1898-1900. Vice President of AIME, 1901-02.

Samuel Stewart Fowler, 1900-1901

B. 1860, Mount Vernon, NY. d. 1940. Graduated in engineering from Columbia University. Arrived in British Columbia when mining was in its initial stages. Played an important role in the development of the Kootenay region. Engineer and later general manager of London and B. C. Goldfields Ltd. Developed, equipped, and operated the Ymir gold mine, the Whitewater and Enterprise mines in the Slocan, and the Cascade power plant. "A man of sagacity and sound thinking." **CIM:** Served on Council from 1898; President 1900-01; Chairman of B.C. Branch at Nelson, 1902, 1909, 1911; Honourary Member 1926.

Charles Fergie, 1901-1903

B. 1857, Wigan, Lancashire, England. d. October 19, 1924, Montreal, QC. Came to Canada in 1887 to manage the Intercolonial Coal Mining Co. in Westville, NS; vice president and general manager, 1890-1905. General superintendent of mines, Dominion Coal Co. in Glace Bay. In

1908, retired and moved to Montreal where he was a consulting engineer for Mackenzie and Mann, which later opened the Inverness coal mines. **CIM:** President MSNS, 1898-1900; Institute Founder; President 1901-03; Vice President 1914-17. Member of CSCE, AIME, and North of England Mining and Metallurgical Institute.

Eugene Coste, 1903-1905

B. 1859, Essex County, ON. d. January 23, 1940, Toronto, ON. Educated in Grenoble and Paris, France. Graduated in 1883 from l'École Nationale Supérieure des Mines. Returned to Canada. Employed by the Geological Survey of Canada. Later opened a private practice as a mining engineer. Manager of the Provincial Natural Gas and Fuel Co., Ontario, 1906. Organized the Volcanic Oil and Gas Co. in Kent County. In 1909, a consulting engineer and geologist for the Canadian Pacific Railway. Discovered the Bow Island gas field in southern Alberta. Purchased these holdings, and formed a company which supplied natural gas to Calgary. A pioneer in the development of Canada's petroleum and natural gas industry. **CIM:** Member 1898; Councillor 1899-1900; Vice President 1901-02; President 1903-05. An important player in the guidance of the Institute in its early years. First Canadian-born CIM President.

George Robert Smith, 1905-1907

B. February 17, 1860, Newark, NJ. d. February 20, 1922. Operated mica, graphite, and phosphate properties near Buckingham, QC. Worked for Ingersoll Rock and Drill Co. Vice president and mine manager of the Bell Asbestos mine in Thetford Mines, QC. Member of the Legislative Assembly for Megantic, QC, 1897-1907, and 1911-22. Major in the 11th Regiment of Hussars, Canadian Militia. Keenly interested in gardening, farming, and horses. **CIM:** Institute Founder; Vice President 1903-04; President 1905-07. His son, George W. Smith, was President Elect for 1970.

Frederic Keffer, 1907-1908

B. September 21, 1861, Philadelphia, PA. d. March 12, 1947, Spokane, WA. Graduated in mining engineering from Ohio State University, 1882. Engaged in railway construction in Ohio and Pennsylvania, and electrical engineering in Columbus, OH, then accepted a position at Ohio State University as assistant professor of chemistry until 1892. Worked as a chemist for a natural soda plant in Wyoming, supervised a gold mine in Querobabi, Mexico, and operated an electrochemical plant in Elizabeth, NJ. Came to Canada in 1896. Worked for the Boundary Mining Syndicate, and later the British Columbia Copper Co., Greenwood, BC, until that plant closed because of the war. In 1915, was a private consultant in Spokane, WA, and supervised the opening and operation of several copper, gold and silver mines, as well as a smelter. The most important was at Copper Mountain, near Princeton, BC, later operated by Granby Consolidated Mining, Smelting, and Power Co. Ltd. He was one of the most respected engineers in the Pacific Northwest. **CIM:** Member 1901; Councillor 1902-3, 1908-11; Vice President 1905-06; President 1907-08; Chairman of the Western Branch 1908; Honourary Life Member 1925. Member of AIME. Fellow of the Royal Society of Arts, London, 1938.

Willet Green Miller, 1908-1910

B. July 19, 1866, Charlotteville Township, ON. d. February 4, 1925, Toronto, ON. Studied at Harvard, Chicago, and Heidelberg Universities. Member of the Geological Survey of Canada. Professor at Queen's University, awarded an LLD Provincial geologist of the Ontario Department of Mines 1902-25. Identified the silver deposits at Cobalt, ON. During the First World War was the Canadian representative on the Imperial Resources Bureau in London, and later, a commissioner who reported on the nickel industry in Ontario. Leading authority on the Precambrian Shield, and advisor on mineral development, particularly in Ontario. "A man of sterling character." **CIM:** Charter member of the Ontario Mining Institute; CIM Member 1898, Vice President 1907-08; President 1908-10. First Canadian to win the Institution of Mining and Metallurgy (UK) gold medal. President of the Geological Section of the British Association for the Advancement of Science. Inducted into the Canadian Mining Hall of Fame in 1955. Miller Memorial Tablets were presented by CIM to several Canadian universities: "his work on the Precambrian is a monument of inspitation to all students of geology."

Frank Dawson Adams, 1910-1912

B. September 17, 1859, Montreal, QC. d. December 26, 1942, Montreal, QC. Began as an assistant to the local pharmacist. Graduated in 1878 from McGill University in geology. Worked for the Geological Survey of Canada as assistant chemist and lithologist. A pioneer in petrography. Studied at Yale, Johns Hopkins, Zurich, Heidelberg (PhD). Honourary degrees from Toronto, Queen's, and Bishop's Universities. Appointed Logan professor of geology and palaeontology at McGill University. Dean of faculty of applied science and later vice-principal emeritus. Served in the Khaki University, on the National Research Council of Canada, and on the Commission of Conservation. Carried out geological field studies for 25 years on the Grenville province of eastern Ontario, and Quebec's Monteregian Hills. Author of The Birth and Development of the Geological Sciences, published in 1938. An outstanding scholar, and a gifted teacher. **CIM:** Member 1898; President 1910-12. Invited notable American scientists to the Annual General Meetings, and promoted the CIM among his students. Honourary Member of the Institution of Mining and Metallurgy (UK), and AIME. Fellow and president (1914) of the Royal Society of Canada. Fellow of the Royal Society of London. Recipient of Wollaston Medal (1939), and numerous other awards and distinctions. His wife was the founder of the Women's Association of the Montreal Branch of CIM in 1921.

Alfred Ernest Barlow, 1912-1914

B. June 17, 1861, Montreal, QC. d. May 29, 1914, QC. Graduated 1883 with honours in natural science from McGill University. Honourary lecturer in geology at McGill University. Joined the staff of the Geological Survey of Canada in 1883. An assistant to R.W. Ells. Worked in the Shickshock Mountains, Gaspé, in 1883, and in the Cobequid Mountains in 1884. Studied the Archean rocks in the Lake of the Woods region, then in the Sudbury, Nipissing and Temiskaming districts. First geologist to identify the Huronian belt as a promising location for mineral deposits. Wrote a definitive report on the nickel and copper deposits of Sudbury. Collaborated with F.D. Adams on a geological study of the Grenville Province, and the Haliburton-Hastings district of eastern Ontario. In 1900, was appointed Geological Survey of Canada petrographer. Examined the zinc mines of Vancouver Island and the West Coast as part of a federal commission in 1905, then investigated iron ore deposits at Lake Timagami. Left the Geological Survey of Canada to become a consulting engineer for a German syndicate in the Gowganda district of Ontario, then in 1910, was in private practice in Montreal. He and his wife were ill-fated passengers on the Empress of Ireland which sank in the St. Lawrence River near Rimouski in 1914. **CIM:** Vice President 1908-12; President 1912-14; Councillor for several terms; Secretary of the Montreal Branch 1910. Helped to plan the successful International Geological Congress. The Barlow Memorial Medal was established in his memory for the best paper in economic geology published in the *Bulletin*; awarded yearly since 1926, it now extends to papers published in *Exploration and Mining Geology*.

George G. S. Lindsey, 1914-1916

B. 1860, Toronto, ON. d. May 27, 1920, Toronto, ON. Graduated from University of Toronto and Osgoode Hall. Member of the Bar in Ontario, 1886. General manager and later president of Crow's Nest Pass Coal Co., Fernie, BC. Representative for British mining interests in Canada, and spent some time in China, 1915-16, where he helped draft a mining act for the Chinese Republic. A writer and an outstanding athlete, particularly cricket. **CIM:** Member, 1903; Councillor 1910; Vice President 1912-14; President 1914-15, 1915-16. Responsible for initiating negotiations with MSNS for affiliation. Executive member of the 12th International Geological Congress in 1913. Initiated a change in the publication and format of the *Bulletin*. A great entertainer at the AGM.

Arthur Augustus Cole, 1916-1918

B. 1871, Montreal, QC. d. October 25, 1959, Ringwood, ON. Graduated with an MA from McGill University, 1897. Employed by the Geological Survey of Canada as an assistant to Albert Peter Low in the St. Maurice area, and Alfred E. Barlow in the Cobalt region. Mapped Ontario's first gold mining area in Hastings. In 1897-1905, was chief chemist and assayer, then supervisor for the War Eagle and Center Star mines at Rossland, BC. In 1906, was mining engineer for the Temiskaming and Northern Ontario Railway Co., and recommended that a rail line be built into the Porcupine area. Also suggested a railway extension through Kirkland Lake to Noranda, QC. From 1921 to 1950, was manager of the Temiskaming Testing Lab at Cobalt (later purchased by the Ontario Department of Mines). Pursued private study on deep mining. Reputedly had "the world's best" collection of mineral specimens from the Cobalt camp. **CIM:** Member 1899; Chairman of the Cobalt Branch; District Councillor 1908-12; Vice President 1912-13, 1915-16; President 1916-18; Life Member 1949; Charter member of Fifty-year Club in 1958.

Donaldson Bogart Dowling, 1918-1919

B. November 5, 1858, Camden Township, ON. d. May 26, 1926, Ottawa, ON. Graduated 1883 in applied science from McGill University with a gold medal. Honourary DSc from McGill, 1921. Joined Geological Survey of Canada in 1884, and was an officer for 34 years. Assistant to R.G. McConnell, G.M. Dawson, and J.B. Tyrell. Studied the coal, petroleum, and natural gas deposits of the Great Plains. A foremost authority on coal. Designed and implemented an artesian water system for irrigation in southern Alberta in 1915. Author of numerous articles and papers. An editor of *The Coal Resources of the World*, issued by the 12th International Geological Congress in 1913. Member of the Dominion Fuel Board, and the Geographic Board of Canada. Fellow of the Royal Society of Canada. Well known in Ottawa as a lecturer in popular science. Enjoyed music, and painted landscapes in oils and watercolours. **CIM:** Member 1904; Councillor 1912, 1914; Vice President 1916-18; President 1918-19. CIM dedicated a cairn and bronze plaque in his memory in Banff in 1933 in recognition of his early work in the Cascade Coal Basin.

Donald Hugh McDougall, 1919-1920

B. September 15, 1879, Sydney, NS. d. 1957, NS. Graduated from Dalhousie University. Awarded an LLD from St. Francis Xavier University. First worked for the Dominion Coal Co. in Glace Bay. Joined Dominion Iron and Steel Co. as manager of Wabana iron mines in 1905. Then was general superintendent, general manager (1912) and president (1919) of Nova Scotia Steel and Coal Co. In 1921, was vice-president of the British Empire Steel Co. Consulting mining engineer in Montreal from 1924. Helped develop many mining projects in North America. Colonel during World War I. During World War II, was an advisor to the Department of National Defence, and chaired the joint Central Committee of the Engineering Institute of Canada. "A Gaelic scholar of local renown." **CIM:** MSNS President 1915-18; CIM President 1919-20; Councillor of both organizations for many years; Chairman of Executive Committee of Council for 10 years; Honourary Member of MSNS. Member of Engineering Institute of Canada, AIME, American Iron and Steel Institute, and Steel Institute of Scotland. Received the Queen's Coronation Medal. Instrumental in securing affiliation of MSNS with the CIM in 1918. An uncle to 1989-90 CIM President John Laffin.

Orton Edward Simpson Whiteside, 1920-1921

B. 1870, Metcalfe, ON. d. August 4, 1942, Vancouver, BC. Graduated from McGill University with a BASc in natural science, and MSc in 1900. Mining engineer for H.W. McNeil Co. at the Anthracite and Canmore colleries, later appointed general superintendent. In 1904, was general manager of West Canadian Colleries Ltd. at Blairmore. From 1908-1935 was general manager of International Coal and Coke Co. at Coleman, 1908-1935. General manager of Coleman Light and Water Co., 1924-1935. An authority on colliery engineering practice and management. "A man of unusual charm and friendly disposition." **CIM:** Member 1898; Councillor 1910-11, 1914-15, 1931-32; President 1920-21; one of the organizers of the Rocky Mountain Branch. Introduced the concept of geographical Divisions (later called Districts) in 1921. Past president and director of the Western Coal Operators' Association which he helped organize.

C.V. Corless, 1921-1922

B. July 30, 1868, Oxford County, ON. d. February 7, 1953, Tillsonburg, ON. First studied to be a teacher. Graduated from McGill University in 1902 with a BSc in mining and metallurgy. Won British Association Medal and Dawson Fellowship in mining. Earned MSc in 1903. Awarded LLD by McGill University and Queen's University. Worked as general manager of Mond Nickel Co., 1903-1918, then director until 1928. Retained as a consultant after Mond Nickel merged with Inco. Had wide-ranging interests; wrote, and lectured on educational matters and industrial relations. **CIM:** President's Gold Medal for best student essay 1901. Member 1903; Councillor 1923-25. President 1921-22. President of Ontario Mining Association. Member of Dominion and Ontario Research Councils, Institution of Mining and Metallurgy (UK), and Royal Canadian Institute.

William R. Wilson, 1922-1923

B. Derbyshire, England. d. March 24, 1937. Worked as an apprentice to his father, an eminent mining engineer, before immigrating to Canada. Returned to England and obtained a certificate as colliery manager. Worked in coal mines in South Wales, Derbyshire, Durham, and finally as manager of a colliery in North Staffordshire. Went to the US in 1883 as consulting engineer for the Economic Society of Pittsburg, PA. Appointed general manager of Crow's Nest Pass Coal Co. in 1900. From 1901 to 1903, was general manager of Veereeniging Estates Ltd., in South Africa. Returned to Canada in 1904 as a consulting mining engineer in Toronto. In 1911, again became general manager of Crow's Nest Pass Coal Co. to deal with mining and labour problems. Appointed director, then president in 1920. Director of Premier Gold Mining Co., and involved in the development of its gold mine near Portland, BC. Designed and built the first steel tipples and headframes of their kind in North America, the first two-car, pendulum counter balance side-dump, and the 'Wilson' type ventilating fan. "A man of broad outlook and sympathies, and a fluent speaker, whether the topic be of technical or general interest." **CIM:** Member 1901; one of the organizers of the Rocky Mountain Branch in 1912, and its first Chairman; Councillor 1918-19; Vice President 1921-22; President 1922-23.

John Alexander Dresser, 1923-1924

B. June 27, 1866, Richmond, QC. d. July 5, 1954, Montreal, QC. Graduated from McGill University in 1897 with an MA in geology. Post-graduate studies at Harvard. Awarded LLD by McGill University in 1933. Principal of St. Francis College, Richmond, in 1895, then lecturer in geology, McGill University, 1903-09. Officer of the Geological Survey of Canada, 1909-1911. Geologist for the Algoma Central and Hudson Bay Railway, 1911. Geologist and manager of Lands Department, Lake Superior Corp. (Algoma Railways), 1911-15. Consulting geologist in Montreal from 1915, where he acted as development geologist for the Canadian Pacific Railway, 1922-32. Directing geologist of Quebec Department of Mines, 1929-39. Author of numerous geological reports and papers. **CIM:** Member 1902; Councillor 1910-11, 1916-17; Vice President 1918-19; President 1923-24; Honourary Member 1951. Member of AIME, Institution of Mining and Metallurgy (UK) , Fellow of the Royal Society of Canada, Geological Society of America, Society of Economic Geologists. Distinguished geologist and leading authority on the geology of Quebec's Eastern Townships. The mineral dresserite is named after him.

Robert Charles Wallace, 1924-1925

B. 1881, Orkney Islands, Scotland. d. January 29, 1955, Kingston, ON. MA from Edinburgh University in 1901 and BSc in 1907. PhD from Gottingen University in 1909. DSc from Edinburgh University in 1912.

Recipient of 20 honourary degrees. Demonstrator in crystallography at St. Andrew's University. Came to Canada in 1910. Lectured in geology and mineralogy at University of Manitoba, and head of department, 1912- 1928. Served as commissioner of northern Manitoba, 1918-21, and commissioner of mines, 1927. President of University of Alberta, 1928-1935. Principal and vice chancellor of Queen's University, 1936-51. One of three Canadian delegates in creation of UNESCO. Chairman of the Ontario Research Commission. President of Research Council of Ontario. President of Royal Society of Canada. Member of National Research Council. After retirement in 1951, he was director of the Arctic Institute. **CIM:** Member 1913; helped organize Winnipeg Branch in 1918; Councillor 1919-20, 1931-33; Vice President 1921-23; President 1924-25; Life Member 1944. A practical man with a lively and active interest in the teaching profession, and the education of young people. He never lost interest in geology, even though his administrative duties made it difficult to practice as a geologist.

Thomas J. Brown, 1925-1926

B. January 21, 1867, Sydney Mines, NS. d. 1926, NS. Worked his way up from an office boy to manager of the General Mining Association at Sydney Mines. Clerk and telegraph operator at Lingan Colliery. Paymaster, then manager of the Victoria Mines, 1873-1896. Manager of Caledonia mine, 1897-1899. Assistant general manager of Dominion Coal Co. in 1900. Manager, then general superintendent of Nova Scotia Steel and Coal Co., responsible for colleries at Sydney Mines, steel plants at Sydney Mines and Trenton, iron works at Ferrona, and iron mines at Wabana, NF, 1901-21. Manager of Inverness Railway and Coal Co., then retired as a consulting mining engineer. In 1923, was Nova Scotia deputy minister of Public Works and Mines. A noted raconteur, especially about the Scots of Cape Breton. Possessed a magnificent and rare library rich in Canadiana. **CIM:** MSNS President 1910-12; CIM President 1925-26. Member of Engineering Institute of Canada, American Society of Civil Engineers, Society of Professional Engineers of NS. Member of Board of Examiners which granted certificates of proficiency to coal mining officials in Nova Scotia.

Lewis Stockett, 1926-1927

B. March 31, 1861, Ashland, PA. d. December 19, 1943, Vancouver, BC. Devoted his entire professional life to the coal industry. First worked as a mining engineer for Lehigh Valley Coal Co., 1875-81, then for Westmoreland Coal Co. near Pittsburg, 1881-84. Superintendent of Park Collieries, 1884-86, and chief engineer at Consolidated Coal Co., St. Louis,1886-91. Secretary-treasurer of Wabush Coal Co., Indiana, until 1895, then appointed mining engineer for the Great Northern Railroad, general manager of Great Northern coal mines at Great Falls, MT, 1897-1904. Came to Canada in 1905. Manager of Canadian Pacific Railway's Bankhead coal mines near Banff, AB, later that company's mines at Hosmer, BC. General superintendent of the Coal Mine Branch of the Canadian Pacific Railway's Department of Natural Resources, 1912-1929. **CIM:** Member 1905; Councillor 1913-18, 1920-22; President 1926-27. Member of Engineering Institute of Canada, and Association of Professional Engineers of Alberta. President of the Western Coal Operators Association, 1909-16. Legion of Honour member of AIME.

John L. Agnew, 1927-1928

B. July 28, 1884, Pittsburg, PA. d. July 8, 1931, Copper Cliffs, ON. Joined the Canadian Copper Co. at Copper Cliff, ON in 1904, becoming smelter superintendent, then general superintendent. Held executive positions in various companies which existed prior to their 1929 consolidation as Inco, of which he became vice president. Directed the opening and development of the Frood mine, the construction of Copper Cliff's new smelter, hydroelectric projects on the Spanish River, and the expansion of the refinery at Port Colbourne. **CIM:** Vice President 1926-1927; President 1927-28. President of Ontario Mining Association. Vice President of AIME, 1925-27.

Robert Holden (Pat) Stewart, 1928-1929

B. 1874, New Brunswick. d. December 23, 1952, Vancouver, BC. Obtained BSc from McGill University in 1896. Worked at the Le Roi mine in Rossland. Later was superintendent of War Eagle and Center Star mines. In 1911, succeeded W.H. Aldridge as general manager of Cominco. Remained there until his retirement in 1918. Adopted the 'French' process for zinc treatment at Trail, which led to the first production of electrolytic zinc in Canada. Then opened a mining consultant practice in Vancouver. Recommended that the Sullivan mine be acquired by Cominco, then directed long-term planning and operational revamping of the mine. As consultant for Texas Gulf Sulphur, he applied his initiative and skill to the extraction of sulfur by liquefaction. Consultant for Mond Nickel, Newmont Mining Corp., Government of Canada, and Canadian Exploration (subsidiary of Placer Development). **CIM:** Member 1907; Councillor 1908-09, 1918-20, 1929-31; Vice President 1912-13; President 1928-29; Life Member 1945. Active in several engineering and professional organizations. Member of the mining committee of the Honourary Advisory Council for Scientific and Industrial Research in Canada. Was affectionately called the "Dean of the profession of mining engineering" in British Columbia.

Norman Richard Fisher, 1929-1930

B. Christchurch, New Zealand. d. June 15, 1961, Montreal, QC. Graduated from Otago University School of Mines, New Zealand, with BSc in mining engineering. Worked in New Zealand, England, and the United States in several supervisory positions. Moved to Canada in 1907. Was a mining and consulting engineer with Temiskaming Mining Co. Ltd., Cobalt; Wright-Hargreaves mine, Kirkland Lake; Consolidated Asbestos Co. Ltd., Thetford Mines; Shield Development Co. Ltd., and Moss Mines Ltd. in Ontario. In 1918, was appointed manager of the Jacobs Asbestos Mining Co., Thetford Mines. Author of numerous publications related to mining. An enthusiastic sportsman who enjoyed cricket, soccer, golf, and curling. **CIM:** Member 1908; Life Member 1921; President 1929-30; Member of Fifty-Year Club 1958. In 1924, he was Canada's official delegate to the 1st Empire Mining Congress in London, England. Member of main committee of the 2nd Empire Mining Congress in Canada in 1927, and vice president of the 3rd Empire Mining Conference in Africa in 1930. Member of the Institution of Mining and Metallurgy (UK). A CIM member for 53 years, he once held the distinction of being the oldest living past president, and the only one originating from New Zealand.

Justin S. DeLury, 1930-1931

B. 1894, Manilla, ON. d. October 1968. Graduated with a BA in mineralogy and chemistry from the University of Toronto in 1905. Assistant in Department of Mineralogy, University of Toronto, then lecturer, University of Idaho. Appointed associate professor in 1911, but left this post to go ranching with his brother in the Peace River country of Alberta for four years. In 1915, returned to teaching, was appointed assistant professor of geology at University of Manitoba. Took a sabbatical in order to study at the University of Minnesota, obtaining a PhD. Professor of geology at the University of Saskatchewan in 1927. Appointed head of the Geology Department at University of Manitoba, and commissioner of mines for that province, in 1928. Appointed provincial geologist in 1929, when natural resources were transferred from federal to provincial jurisdiction. Resigned in 1944, and returned to Ontario. Devoted his time to the study of Precambrian structural problems, on which he had already authored a number of papers. **CIM:** Member 1918; Councillor 1920-22, 1929-30; Vice President 1925-26; President 1930-31; Life Member, 1944; Member of the Fifty-Year Club 1968. Fellow of the Geological Society of American, and Fellow of the Royal Society of Canada, of which he was president of Section IV (Geology) in 1944. "His friendly manner and quiet sense of humour endeared him to his friends, and he was always fondly remembered by his many students."

Francis William Gray, 1931-1932

B. April 15, 1877, Yorkshire, England. d. June 10, 1958, Victoria, BC. Graduated in mining engineering from St. George's Technical College, Sheffield (later Sheffield University), Awarded LLD by Dalhousie University in 1937. Worked at Wharncliffe Silkstone Colleries at Tankersley, Yorkshire, the last four years as chief surveyor. Arrived in Canada in 1904. Worked for Dominion Coal Co. at Glace Bay, NS, and associated companies, from 1904-1945, except, in 1919-21, when he was editor of the *Canadian Mining Journal* and *Iron and Steel of Canada*. Assistant general manager of the Dominion Steel and Coal Corp. Ltd., 1928-45. In 1938, prepared a report on the operations at Canadian Colleries (Dunsmuir), Vancouver Island, for a Royal Commission. In 1939, investigated labour difficulties in that region. Credited for introducting, in 1907, the "Draeger" self-contained oxygen breathing apparatus to Nova Scotia coal mines. Prolific author of papers on Nova Scotia coal fields. **CIM:** MSNS member 1906, Councillor for numerous terms, and President 1927-28; CIM Member 1909; served five terms as Councillor between 1911 and 1947; Vice President 1928-29; President 1931-32. Recipient of the Barlow Memorial Medal in 1927. Member of the Institution of Mining and Metallurgy (UK), Engineering Institute of

Canada, the Economic Council of Nova Scotia, and the Committee of the International Labour Organization. An excellent artist and sculptor, created a plaque of CIM President D.B. Dowling, a replica of which hung for many years at CIM. His son, R. Heath Gray, was CIM President in 1982-83, making them the only father and son combination in CIM history.

John Andrew Allan, 1932-1933

B. July 31, 1884, Aubrey, QC. d. June 6, 1955, Edmonton, AB. Graduated from McGill University with an MSc in geology in 1908. Earned a PhD at Massachusetts Institute of Technology in 1912. Established the Department of Geology and Mineralogy at University of Alberta in 1912, becoming professor and head in 1913. Retired in 1949 as professor emeritus. Spent field season of 1906-18 with Geological Survey of Canada. In1919, organized what is now the Research Council of Alberta. Geological consultant for the Bow River power dams for Calgary Power (TransAlta Utilities), Skiff oil discovery in 1927, and Kinsella gas discovery in 1930. Spent his early career mapping coal seams in the Kananaskis Range, Alberta. Mount Allan, the site of the downhill skiing events for the 1988 Winter Olympics (Nakiska), was named in his honour. An excellent teacher known for his large collection of minerals and fossils which made the University of Alberta geological museum one of the best in western Canada. **CIM:** Member 1913; Councillor 1919-23; Vice President 1923-24; President 1932-33. Fellow of the Royal Society of Canada. Charter member of the Alberta Society of Petroleum Geologists (1927). President of the Association of Professional Engineers of Alberta in 1930.

Balmer Neilly, 1933-1934

B. July 26, 1883, Bradford, ON. d. February 3, 1956, Toronto, ON. Graduated from the University of Toronto in 1908 with BASc in mining engineering and ME in 1915. Awarded LLD in 1942. Worked at John Black Mining Co., Crystal, CO. In 1909-1919, employed at the Silver Queen, Beaver, Black, and Wyandoh mines in Cobalt. Manager of Penn-Canadian Mines Ltd. in Cobalt, 1912-13. Assistant to the president of McIntrye Porcupine Mines in 1924, becoming president and managing director in 1946. Director of several mining companies. **CIM:** Member 1908; won the student's essay competition the same year; Councillor 1927-28, 1934-35; Vice

President 1918-20; President 1933-34. Member of The Institution of Mining and Metallurgy (UK). Helped form the Temiskaming Mine Managers' Association, and in 1920, the Ontario Mining Association. Member of Board of Governors of University of Toronto. In 1939-49, chaired the Governor's Committee on Connaught Laboratories, which set records for the production of penicillin and blood plasma during World War II. Active in setting up McIntrye Research Foundation, which helped to combat the effects of silicosis in hard-rock miners.

Selwyn Gwillym Blaylock, 1934-1935

B. February 18, 1879, Paspebiac, QC. d. November 19, 1945, Trail, BC. Attended Bishop's College School, Lennoxville, then McGill University, graduating in 1899 with a BSc, LLD from McGill University in 1929, and University of Alberta in 1930. Moved to Trail and worked as an assayer for the Canadian Smelting Works. In 1901, was chief chemist with Consolidated Mining and Smelting Co. (Cominco), becoming assistant general manager in 1910, general manager in 1919, president in 1943, and chairman in 1945. A distinguished metallurgist, he was instrumental in the development of the electrolytic zinc process at Trail in 1916. Although he earned considerable stature in industry, he never lost sight of the welfare of people in isolated towns who depended on the mining industry for their livelihood. **CIM:** Secretary-Treasurer of the Rossland and Nelson Branches 1898-1901; Member 1907; Councillor 1910-11, 1926-27; President 1934-35; Honourary Member, 1944; recepient of Inco Medal in 1935. Director of AIME 1936-39. Received the McCharles Prize from University of Toronto in 1924, James Douglas Gold Medal (AIME) in 1928. The Selwyn G. Blaylock Medal was named in his honour. Inducted into Canadian Mining Hall of Fame in 1990.

James Gordon Ross, 1935-1936

B. 1879, Embro, ON. d. Spring of 1968, Montebello, QC. Graduated with a BSc in mining engineering from McGill University in 1903. Worked during his undergraduate summers in the mines at Rossland and Fernie, BC, and Glace Bay, NS. Starting in 1903, was a maintenance-of-way engineer with the Grand Trunk Railway. Spent a year prospecting in Cobalt. Worked for two years as engineer and superintendent on the Hudson River Tunnel Project in New York. In 1907, was superintendent of the Mond Nickel Co. Worthington mine, and in 1908-09, travelled through Europe, Asia, Africa, Australia, and New Zealand to examine mining properties. In 1911 became consulting engineer for the Milton Hersey Co. of Montreal. During World War I, served with 1st Canadian Contingent as Lieutenant, 13th Battalion, Royal Highlanders, earning rank of Captain. Wounded in Second Battle of Ypres in 1915, recovered, then severely injured at Festubert. After the war, returned to the Milton Hersey Co. Then became a consulting engineer to Black Lake Chrome and Asbestos Corp. By 1929, was general manager of Asbestos Corp. Ltd. **CIM:** Member 1911; Councillor for 13 years; Chairman of Montreal Branch 1920; Vice-President 1928-29, 1932-33; President 1935-36; awarded Randolph Bruce Gold Medal in 1935. Member of the Anti-Tuberculosis Association, and Asbestos Producers' Association. An uncle to Charles Bruce Ross, CIM President in 1986-87.

George Edwards Cole, 1936-1937

B. December 2, 1879, Aylmer, QC. d. December 11, 1957, Winnipeg, MB. Graduated from McGill University with a BSc in 1906. After working in BC, and Ontario's Gowganda District, taught at the Haileybury School of Mines. In World War I, was a Lieutenant-Colonel with the Canadian Expeditionary Force and the Tunnelling Company of the Canadian Engineers. In 1921, became inspector of mines for Ontario. In 1928, was chief inspector of mines for Manitoba Department of Mines and Mineral Resources, and its director 1930-45. **CIM:** Chairman of Winnipeg Branch 1930 and 1944; Vice President 1933-34; President 1936-37. The Past Presidents' Memorial Medal was established in his honour in 1959.

Michael (Mike) Dwyer, 1937-1938

B. February 4, 1877, Ireland. d. December 28, 1953, New Glasgow, NS. Came to Canada at age seven. Superintendent for the Nova Scotia Steel Co. at Sydney Mines, later managed Indian Cove Coal Co. Elected member of the Nova Scotia Legislative Assembly in 1933, and was minister of Mines and Public Works in 1933-39. Resigned in 1939 to be president and general manager of the Nova Scotia Steel and Coal Co. at Trenton. Elected mayor of Sydney Mines in 1926, and mayor of New Glasgow in 1949-50. **CIM:** Member 1910; MSNS President 1931-32; CIM President 1937-38. Member of Professional Engineers of Nova Scotia, and Engineering Institute of Canada.

Intervened in many disputes between miners and operators in colliery towns. Perhaps best known for his role during the collapse of the Moose River gold mine, Halifax County, in 1936, in which he personally supervised the eleven-day rescue efforts.

Everett Alfred Collins, 1938-1939

B. January 19, 1879, Smithfield, ON. d. February 6, 1952, Copper Cliff, ON. Graduated from Queen's University in 1905 with a BSc in mining and metallurgy. As a student worked for Canadian Copper Co. as a construction clerk. After graduation prospected in the Cobalt district, later worked as an assayer. In 1906-10, was superintendent of mines for Osceola Lead and Zinc Mining Co. at Webb City, MO. Returned to Ontario, and spent three years consulting on non-metallic mineral deposits. In 1913-18, was inspector of mines for Ontario. Joined Inco as safety engineer, and in 1920 became assistant to the general manager. In 1929, was elected mayor of Copper Cliff. A distinguished mine executive, concerned private citizen, active in community affairs, and ever conscious of the welfare of those under his jurisdiction. **CIM:** Member 1910; Councillor 1932-36, 1939-42, 1944-46; Vice President, District 3, 1936-37; President 1938-39. Founded and endowed the Medal for Bravery in 1935, often called the Institute's Victoria Cross. Member of Board of Trustees, Queen's University.

Benjamin Leonard Thorne, 1939-1940

B. 1871, Holland Landing, ON. d. March 23, 1944, Calgary AB. Educated in Toronto. First worked in the Canadian Pacific Railway's general freight department. In 1894-97, was assistant Ontario land surveyor in the Nipissing district, then in Lake of the Woods. Carried out mineral investigations in Rainy Lake district, and supervised exploration between Shoal Lake and Echo Bay for Great Granite Gold Mining and Milling Co. Worked at Nina mine in Upper Whitefish Bay. In 1901, engaged by a New England syndicate in Colorado, Arizona, and California. Returned to Canada in 1902. Prospected for iron ore deposits in the Atikokan-Steep Rock Lake area and supervised diamond drilling and magnetometric surveys of properties in Sudbury. In 1903, investigated McInnes coal claims in the Crowsnest Pass for Canadian Pacific Railway. In 1912-13, in association with J.C. Gwillim, examined railway's mineral holdings in Alberta and Saskatchewan, and in 1913, was appointed mining engineer for the Canadian Pacific Railway's Department of Natural Resources. Served as engineer of construction and underground development at Bankhead, AB, and later Hosmer, BC. Moved to Calgary in 1914 and managed CPR's coal, petroleum, and natural gas rights, until he retired in 1940. He was a leading figure in Alberta's coal and oil industry. **CIM:** Member 1913; Councillor 1925-26; Vice President 1933-34; active in the Rocky Mountain Branch; President 1939-40. President of the Alberta Petroleum Association 1943-44.

George Cecil Bateman, 1940-1941

B. December 25, 1882, Sutton, ON. d. February 4, 1963, Oche Rios, Jamaica. Graduated from Queen's University in 1905 with a BSc in mining engineering. Worked as engineer, superintendent, and manager of various mining properties in North America. In 1915-24, was general manager of La Rose Mines in Cobalt. Named metals controller for Canada in the Department of Munitions and Supply during World War II. In 1944, was transferred to Washington, DC, as Canada's deputy member of the Combined Production and Resources Board. In 1946, was appointed to the Atomic Energy Control Board. Wrote extensively on technical matters. Held directorships in the Algoma Steel Co., Algoma Ore Properties, Canada Steamship Lines, Eastern Trust Co., and White Pass and Yukon Corp. In 1922-1940, was secretary-treasurer of the Ontario Mining Association. **CIM:** Member 1919; President 1940-41; Honourary Member 1948. A leading Canadian who helped organize and conduct Canada's war effort at the same time that the mining industry was at its peak of wartime production.

Wilbert George McBride,1941-1942

B. February 8, 1879, Inglewood, ON. d. August 22, 1943, Dorval Island, QC. Graduated with a BSc in mining engineering from McGill University in 1902. Went to Bisbee, AZ, to work as chief engineer at the Copper Queen mine. In 1907, became superintendent of the Sierra de Cobre Mines in Cananea, Mexico (a subsidiary of the Copper Queen). A few months later, became general superintendent of the Great Western Copper Co. at Courtland, AZ. In 1916, was assistant general manager of the Morenci Copper Co., then general manager of the Old Dominion Co. at Globe, AZ. In 1927, he was appointed Macdonald professor and head of the Department of Mining Engineering at McGill University. Was later a mining consultant in the Montreal area. **CIM:** Member 1927; Chairman of the

Montreal Branch; served on several Institute committees; Councillor 1928-34, 1935-39; Vice President 1934-35; President 1941-42. As President, he visited and addressed almost every CIM Branch. As Chairman of the Executive Committee in 1942, he was chiefly responsible for by-law amendments on membership qualifications and classification. Awarded the Julian C. Smith Medal (Engineering Institute of Canada). Member of AIME, and chairman of its Arizona Section in 1920. Council member of the Institution of Mining and Metallurgy (UK).

Robert A. Bryce, 1942-1944

B. 1881, Toronto, ON. d. October 12, 1963, Toronto, ON. Graduated in 1904 with a BSc in mining engineering from University of Toronto. Awarded an Honourary DSc from the University of Ottawa. First worked in Cananea, Mexico, where he was superintendent of a silver mine. Returned to Ontario, and worked in Cobalt as manager of the Silver Queen, Beaver Consolidated, and Silver Cliff mines. In 1912, opened a successful private practice in Toronto, specializing in mining exploration and development. Was president and managing director of Macassa Gold Mines, Kirkland Lake, which he helped to develop. Directed the start-up of the Renabie mine, and the Bicroft uranium mine. Director of several companies. Chairman of the War Metals Advisory Committee during World War II. President of the Canadian Chamber of Commerce in 1949. Long-time member of the Board of Governors of the University of Toronto. **CIM:** Member 1906; Councillor 1910-13, 1919-21, 1925-26; Vice President 1927-28; President 1942-44; Life Member 1958. In 1957, chaired the General Committee of the 6th Commonwealth Mining and Metallurgical Congress in Vancouver. His wife Edna (Baxter) was president of the Women's Association of the Mining Industry of Canada in 1926-30 and 1934-36.

A. Allen Mackay, 1944-1945

B. July 26, 1885, Gould, QC. d. May 6, 1946. First worked in the gold mines of Idaho Springs, CO. Returned to Canada in 1903. Graduated from Queen's University with a BSc in mining engineering in 1910. Mine captain and superintendent at the Helen mine, Michipicoten, ON. In 1913, mine manager in Sudbury for the Ontario Bureau of Mines. Then opened an iron mine in the Cuyuna range, MN, for a Canadian syndicate, and two manganese properties to supply ore to Algoma Steel Corp. at Sault Ste. Marie. A commissioned officer in World War I with the Royal Canadian Engineers. Afterwards, a consulting engineer for Algoma Central Railway. Prospected in Ontario, British Columbia, and the Yukon. From 1922, rose to vice president and managing director of Aldermac Copper Corp., Francoeur Gold Mines and Lake Geneva Mining Co. In 1945, appointed coordinator of resources development in the Department of Reconstruction. Chaired the Public Investment Committee for the Dominion-Provincial Conference. **CIM:** President 1944-45. Instrumental in organizing the Interprovincial Mining Committee which evolved into the on-going Mines Ministers Conference. While coordinator of resources development, Mackay was known for the phrase "roll back the frontier," a sentiment which directed his many development projects in Canada.

Alan Emerson Cameron, 1945-1946

B. in 1890, London, ON. d. March 7, 1977, Wolfville, NS. Graduated from McGill with a BSc in mining engineering in 1913, and MSc in 1914. Lecturer at the University of Alberta in 1914-16. Spent the 1916-17 field season with the Geological Survey of Canada. In World War I, served with the Royal Canadian Engineers. In 1919, taught at the Khaki University in Yorkshire. In 1926, received a ScD in metallurgy from Massachusetts Institute of Technology, and became professor of metallurgy at the University of Alberta, 1922-37. Honourary degrees from Mount Allison University, Acadia University, and Nova Scotia Technical College. In 1937, appointed deputy minister of mines for Nova Scotia. In 1947-57, was president of Nova Scotia Technical College. Retired in 1957, but continued as a consultant and teacher at St. Mary's University and Dalhousie University until 1967. At age 85, he was still registered as a student at Acadia University pursuing a course during the 1975-76 school term. He was fondly known as "the exploring professor of the University of Alberta." **CIM:** MSNS President 1944-45; CIM President 1945-46; Life Member 1965; helped to strengthen ties between CIM and the MSNS by fostering the establishment of Coal Division, becoming its first Chairman. Also President of Nova Scotia Institute of Science, and Association of Professional Engineers of Nova Scotia. Member of Defense Research Board of Canada, and a governor of Nova Scotia Research Foundation. In 1946, was a member of the Special Mission on Education in Europe. Served as special advisor on education to Indonesia and Burma for the Canadian External Affairs Department in 1956.

Richard J. Ennis, 1946-1947

B. Aspen, CO. d. October 5, 1951. A self-taught mining engineer. In 1911, joined McIntyre Porcupine Mines Ltd., becoming general manager, and a vice president in 1941. Much of the development of improved mining equipment and methods at McIntyre was due to his perserverence. Concerned with safety and well-being of workers. Helped establish the McIntyre Research Foundation to study, treat, and prevent silicosis. **CIM:** Councillor 1934-35; Vice President 1935-36; President 1946-47; recipient of Inco Medal in 1948. Started Branch tour by air, which allowed him to visit almost every Branch in Canada.

Charles Camsell, 1947-1948

B. 1879, Fort Liard, NWT. d. December 19, 1958, Ottawa, ON. Graduated in 1894 from University of Manitoba with a BA in natural sciences. Postgraduate work at Queen's University, Massachusetts Institute of Technology, and Harvard University, Awarded LLD by three Canadian universities. An explorer, geographer, and geologist in the north. Officer of the Geological Survey of Canada, 1904-1920. Appointed deputy minister of the Department of Mines in 1920, and deputy minister of Department of Mines and Resources in 1935. Organized the Dominion Fuel Board in 1922, and was chairman for 20 years. Member of the National Research Council, 1921-36. Chairman of the Canadian Section of the World Power Conference from 1925. One of the commissioners in 1939 who studied waste gas conservation in Turner Valley, AB. Chairman of the Matamek Conference on Biological Cycles. Head of the Canadian delegation at the Imperial Conferences in London in 1923 and 1930, and in Ottawa in 1932. Head of the Canadian delegation at the Empire Mining and Metallurgical Congress in South Africa in 1930. Served on Council of NWT in 1921, and was commissioner in 1935-46. Founded the Charles Camsell Hospital for tuberculosis in Edmonton. **CIM:** Chairman of British Columbia Branch 1919; Vice President 1921-22; President 1947-48; Honourary Member 1948. Member of many technical and scientific societies. Director of AIME 1939-44, and Honourary Member in 1947. Fellow of the Royal Society of Canada. Founder and Fellow of the Canadian Geographic Society, and President in 1929-41. In 1935, awarded Order of St. Michael and St. George by King George V. Holder of numerous other distinctions.

Randolph William Diamond, 1948-1949

B. February 26, 1891, Campbellford, ON. d. October 5, 1978, Victoria, BC. Graduated from the University of Toronto in 1913 with a BSc in mining and metallurgy. Moved to Anaconda, MT, where he was employed in flotation research for the Anaconda Copper Mining Co. In 1917, moved to Trail, BC, and led research team at Consolidated Mining and Smelting (Cominco) in the development of froth flotation process for the treatment of the complex Sullivan mine ore. This led to one of the first successful large-scale differential flotation operations in the world. Responsible for the Trail chemical fertilizer complex, one of the earliest to combat industrial air pollution by converting sulphur dioxide into a saleable product. Appointed superintendent of concentration in 1929, he rose successively to assistant general manager, and deputy head of the Mines and Exploration Department, and vice president and general manager. In 1953, was executive vice president, western region. Retired in April 1956. **CIM:** Member 1928; Councillor 1936-38; President 1948-49; first recipient of the Selwyn G. Blaylock Medal in 1948. President of the 6th Commonwealth Mining and Metallurgical Conference in 1957. Member of the Institution of Mining and Metallurgy (UK), and the Association of Professional Engineers of BC. Awarded the McCharles Prize (University of Toronto) in 1934. Leonard Medalist of Engineering Institute of Canada (EIC) in 1935. Awarded the Julian C. Smith Medal (EIC) in 1948. First Canadian to win the Metals Institute Platinum Medal, in 1953. Inducted into the Canadian Mining Hall of Fame in 1991.

Frederick Victor Seibert, 1949-1950

B. November 8, 1885, Bruce County, ON. d. March 21, 1966, Victoria, BC. Graduated from the University of Toronto in 1912, with a BASc in civil engineering. A surveyor with the Department of the Interior for 15 years, except for two years in the Royal Air Force during World War I. Was known for his geodetic surveys around Great Slave Lake and in the Mackenzie River basin. In 1922, was engaged by the National Development Bureau and the Natural Resources Intelligence Service, Department of the Interior, to look after mining interests in Manitoba and Saskatchewan, becoming superintendent of Natural Resources and Development, Western Region. In 1940, appointed industrial commissioner, western region, Department of Research and Development of the Canadian National Railway. **CIM:** Member 1926; Secretary, and later Chairman of the Winnipeg Branch; Chairman of Industrial Minerals Division 1946-48; Councillor 1931-33, 1940-44; President 1949-50. As CIM President, he visit 24 of the 27 Branches. Member of Engineering Institute of Canada, and received the Julian C. Smith Medal. A member of the Association of Professional Engineers of Manitoba, president and director of the Manitoba Chamber of Mines, vice chairman of the Industrial Development Board of Manitoba. Member of the Alberta Research Council.

Alphonse-Olivier Dufresne, 1950-1951

B. April 10, 1890, Montreal, QC. d. June 7, 1989, QC. Graduated in 1911 with a BSc in mining engineering from École Polytechnique, and in 1913 with an MSc from McGill University. Received honourary degrees from both universities. First worked in surveying and construction, then in geological and prospecting work in Quebec. In 1914, was appointed inspector of mines and assistant to the superintendent of the Quebec Bureau of Mines. In 1927, was director of the Bureau of Mines, and in 1941-60, served as deputy minister of mines. Taught mineralogy and geology at the School of Forestry Engineering at Université Laval. Lectured in mineral economics at the Quebec Mining School in Val d'Or, QC, which he helped found in 1938. Created the administrative structure within the Quebec government that led to the growth of Quebec's mining industry. His efforts led to changes in mining legislation and safety, and he encouraged the development of mining communities. Fostered the training of mining engineers, geologists, and metallurgists in Quebec through university scholarship programs, the founding of the École des Mines at Université Laval in 1934, and the establishment of a mining and geological engineering program at École Polytechnique. **CIM:** Member 1912; Councillor 1928-30, 1933-37; Vice President 1930-32, 1937-38; President 1950-51; Life Member in 1961; awarded the Institute's Distinguished Service Medal in 1963. The A.O. Dufresne Award was established in 1978 in his honour. Member and President of the Quebec Corporation of Professional Engineers, member of the Association canadienne française pour l'advancement des sciences, Canadian Geographical Society, Geographical Board of Quebec, and the AIME. Inducted into the Canadian Mining Hall of Fame in 1992. The first French Canadian CIM President. His son, geologist Cyrille D. Dufresne, became a Life Member in 1985.

Oliver B. Hopkins, 1951-1952

B. Virginia. d. circa 1976, Florida. Graduated from Johns Hopkins University with a BA in 1909, and a PhD in 1912. From 1912 to 1914, was assistant state geologist for Georgia, and in 1914-19, a member of the United States Geological Survey. Joined Imperial Oil Ltd. in 1919 when it began explorations in Alberta and the Northwest Territories, and was in charge of an exploration party which operated from Athabasca to the Peace River. Worked in Colombia for a year, then returned to become chief geologist for both the International Petroleum Co. and Imperial Oil. In 1933, he was director of International Petroleum. In 1944, he became vice president of Imperial Oil. Was in charge of exploration and development for Imperial Oil during the Leduc discovery in 1947. Was appointed contract director for Imperial Oil's transportation department in 1949, and in the same year, president of the Interprovincial Pipe Line Co. When the federal Petroleum Division of the Department of Trade (later transferred to the Department of Defence Production) was formed in 1951, he became its director. A prolific author of geological literature, producing several books and numerous articles. **CIM:** President 1951-52.

Desmond Fife Kidd, 1952-1953

B. 1905, London, England. d. October 13, 1962, London, England. Came to Canada in 1911. Graduated with a BSc from the University of British Columbia in geological engineering in 1927. Earned a PhD from Princeton University in 1933. Worked as a student assistant for the Geological Survey of Canada in 1925-1930, then as an assistant geologist. Opened a consulting practice in 1935 in Vancouver. Prospected much of the Northwest Territories, northern BC, and the Yukon, often piloting his own plane. In 1941, staked a property near Revelstoke, later mined by Mastodon Zinc Mines Ltd. of which he became vice president and general manager. Also was a director of Highland Bell Mines Ltd. and president of Attwood Copper Mines Ltd. **CIM:** Member 1929; Chairman of Vancouver Branch, and BC Section 1950; Councillor, District 6, 1942-44; Vice President 1950-51; President 1952-53. Actively pursued a campaign to encourage the formation of Branches in BC, as well as the extension of CIM activities in that province. Member of the Geological Association of Canada, Fellow of the Arctic Institute of North America, member of the Society of Economic

Geologists. Chaired the BC resources Conference in 1954. At 47, he was one of the youngest CIM Presidents, and likely holds a record for the number of Branches visited in a single year.

Arthur Edward Flynn, 1953-1954

B. October 20, 1890, London, England. d. February 28, 1956, Halifax, NS. Moved to the United States at an early age, then returned to England, graduating from the Royal School of Mines in 1912. Worked in England, Germany, and Spain. Moved to Cobalt, ON. Worked for Nipissing Mining Co. as mining engineer and surveyor. Designed and supervised construction of the mining laboratories at the Haileybury School of Mines, and was also an instructor. In 1918-1920 worked for Rand Consolidated Mines Ltd., then joined Hardinge Co. of York, PA. In 1921, appointed professor and chair of mining engineering at Nova Scotia Technical College. In 1947, became honourary registrar of the College, and secretary of both Senate and Board of Governors. Was the College's first dean in 1953. Acted as development engineer for Nova Scotia's Department of Natural Resources. Member of the Nova Scotia Advisory Board for Mineral Development, and of the Nova Scotia Research Foundation Associate Committee on Coal. **CIM:** MSNS Member 1921, Councillor for several terms, and President 1939-40; CIM Vice President, District 1, 1941-42; Councillor 1937-39; Chairman of Halifax Branch (MSNS) 1943; Chairman of Coal Division 1948-49; President 1953-54. Council member of Engineering Institute of Canada, awarded its Leonard Prize in 1930. Active in the Association of Professional Engineers of Nova Scotia. Member of AIME and the Institution of Mining and Metallurgy (UK). In 1953, was awarded the Coronation Medal. "Indefatigable in all that concerned the development of the mineral resources of Nova Scotia."

Charles Coombs Huston, 1954-1955

B. 1908, Michigan. Graduated from Michigan College of Mining and Technology with a BSc in 1930. First worked in Chile and Alaska. Came to Canada in 1934 as an assistant manager of Central Patricia Gold Mines Ltd. Held managerial positions with several other companies. During World War II, was successvively supervising engineer for the Allied War Supplies Corporation in Montreal, assistant to the general manager of Scarboro Munitions Plant, in Toronto, and chief mining engineer for a Brazilian project of the Foreign Economic Administration, Washington. Established a mining consultant practice in 1945. **CIM:** Member 1938; Councillor for several years; President 1954-55; awarded the Inco Medal in 1972.

Alfred Lloyd Penhale, 1955-1956

B. October 11, 1901, Sherbrooke, QC. d. January 6, 1997, North Hatley, QC. First worked at a lumber company, then as a reporter for the *Sherbrooke Daily Record*. Joined Asbestos Corp. in 1924, becoming general sales manager in 1937, secretary in 1943, general manager in 1947, president and managing director in 1948, and chairman in 1967. During World War II, was deputy controller for the asbestos industry under the War Time Prices and Trade Board. Awarded honourary doctorates by Laval University and Bishop's University. Director of numerous companies. **CIM:** Member 1937; Chairman of Thetford Mines Branch; Councillor 1949-51; Executive Committee member 1950-51; Vice President, District 2, 1952-53; President 1955-56; recipient of Distinguished Service Medal in 1965. Vice president of Quebec Asbestos Mining Association, director of Canadian Metal Mining Association, director of Canadian Council International Chamber of Commerce, trustee of Bishop's University. The third generation of his family to have been associated with the Institute and the asbestos industry. His grandfather, Mathew Penhale, and father, Col. John J. Penhale, an Institute founder, were associated with the Asbestos Club, a forerunner of the GMAPQ, which merged with the Federated Canadian Mining Institute.

John Fortune Walker, 1956-1957

B. 1893, Binbrook, ON. d. circa 1974, BC. Graduated from the University of British Columbia with a BASc in geological engineering in 1922, and from Princeton University with a PhD in geology in 1924. Worked for the Geological Survey of Canada in 1921-34. Appointed provincial mineralogist for the BC Department of Mines in 1934, and deputy minister of mines in 1937. Organized the BC War Metals Research Board which later became the BC Research Council. **CIM:** Member 1922; Chairman of Vancouver Branch 1929-39; Councillor 1935-40; Chairman of BC Section; Vice President 1941-43; member of CIM War Metals Advisory Commmittee to the Metals Controller 1942-1945; Chairman of Geology Division 1947; President 1956-57.

Horace John Fraser, 1957-1958

B. November 27, 1905, Girven, SK. d. February 2, 1969, Palgrave, ON. Graduated from University of Manitoba with a BSc in 1925, and an MSc in chemistry and geology in 1927. Obtained a PhD in econcmic geology from Harvard University in 1930. Awarded an LLD from Queen's University in 1958. Teacher and researcher for two years before joining Inco in 1932-35. Professor of geology at California Institute of Technology in 1935-42. Commodity specialist, then assistant chief of the Ferro Alloys Branch of the Foreign Economic Administration of the US Government in Washington in 1942-45. Returned to Canada as manager of Falconbridge Nickel Mines Ltd. becoming general manager in 1947, and president in 1957. President and managing director of Ventures Ltd. in 1958. Oversaw the merger of Ventures Ltd. and Falconbridge Nickel Mines as Falconbridge Ltd., which became the second largest nickel producer in the western world. Developed nickel mining and smelting complex in the Dominican Republic, which became Falconbridge Dominicana. Director of numerous companies. Helped found Laurentian University in 1960. **CIM:** Member 1927; Vice President 1952-53; President 1957-58; awarded the Selwyn G. Blaylock Medal in 1964. Member of many professional societies including AIME, the Society of Economic Geologists, and Association of Professional Engineers of Ontario. President of Ontario Mining Association in 1953, and the Mining Association of Canada in 1954-56. Fellow of the Royal Society of Canada, Geological Association of America, and Mineralogical Society of America. Inducted into the Canadian Mining Hall of Fame in 1992. A respected mining industry leader.

David G. Burchell, 1958-1959

B. 1909, in Joggins Mines, NS. d. January 30, 1994, Bras d'Or, NS. Fourth generation of his family to operate coal mines in Nova Scotia. Graduated with BSc from Mount Allison University. Started in the coal mining business in 1936. By the 1950s, was owner and operator of five underground coal mines in Cape Breton: the Toronto, Franklyn, and Cranberry mines in Bras D'Or, the Beaver, and 4-Star near Broughton. Member of the Nova Scotia Labour Relations Board. Canadian delegate to the International Labour Organization in Turkey in 1956, and Geneva in 1959 and 1970. Introduced many innovations in coal mining in Canada, including first coal cleaning plant, first continous underground coal loader, first underground conveyor belt, and first hydraulic long wall roof supports. Prominent in other business activities. Trustee of Cape Breton Miner's Museum. **CIM:** MSNS President 1949-50; Chairman of Coal Division 1950-51; Vice President, District 1, 1950-51; President 1958-59; Life Member 1973; Fifty-Year Club Member 1986; recepient of MSNS Medal in 1977, and CIM Coal Award in 1984. Inducted into the Canadian Mining Hall of Fame in 1995. Was concerned about the social and industrial life of his community and province.

John Campbell Sproule, 1959-1960

B. May 13, 1905, Edmonton, AB. d. May 21, 1970, Jasper, AB. Graduated from University of Alberta with a BSc in 1930, and earned a PhD in geology at the University of Toronto in l935. After gaining wide experience with the Geological Survey of Canada and Imperial Oil, established J. O. Sproule and Associates Ltd. in Calgary, in 1951. Author of numerous papers. Instrumental in bringing the federal government and private industry together to form Panarctic Oils Ltd. **CIM:** Chairman of Petroleum and Natural Gas Division 1953-54; President 1959-60. First Canadian-born president of the American Association of Petroleum Geologists. Awarded the Engineering Institute of Canada's Julian C. Smith Medal in 1966. President of the Association of Professional Engineers of Alberta and recipient of its Centennial Award in l967 for "unusual distinction in the arts and sciences related to the professions of engineering, geology and geophysics".

William Henry Durrell, 1960-1961

B. October 31, 1903, Portage du Fort, QC. d. March 6, 1963, Nassau, Bahamas. Attended Haileybury School of Mines. In 1921, joined Sutcliffe Company Ltd. of New Liskeard, and engaged as consulting engineer for mine development across northern Canada. Was a partner in the firm until 1941. During World War II, was in charge of wartime construction in Labrador, including the building of Goose Bay airport. In 1945, opened his own consulting company. In 1948, became general manager of Labrador Mining and Exploration Co. Ltd., Hollinger Ungava Transport Ltd., Ungava Power Company, and Quebec North Shore and Labrador Railway Co. In 1949 became general manger of Hollinger-Hanna Ltd. (agent for the Iron Ore Company of Canada), and in 1954, vice president. Also vice president and director of the Quebec North Shore and Labrador Railway Co.

CIM: Member 1944; Vice President, District 2, 1955-56; President 1960-61; recipient of the Inco Medal in 1958; served on the Executive Committee of Council, and on special committees. Member of many engineering societies including the Association of Professional Engineers of Ontario, Quebec, and Newfoundland, and the Association of Ontario Land Surveyors. An ardent promoter of the St. Lawrence Seaway project as a transportation corridor for the rich iron ore fields of Quebec and Labrador.

James Buckland Mawdsley, 1961-1962

B. 1894, Italy. d. December 2, 1964. Came to Saskatchewan in 1904 at age nine. In 1913, entered McGill University, but joined the 1st University Company in 1915 during World War I, and was wounded while on active service in France. Transferred to the Royal Flying Corps as a pilot, and wounded again in 1918. Awarded the MBE in recognition of his wartime service with the Princess Patricia's Canadian Light Infantry and the Royal Flying Corps. Returned to Canada, graduated from McGill University in 1921 with a BSc in mining engineering. Then obtained a PhD in mining geology from Princeton University in 1924. Employed by the Geological Survey of Canada in 1924-1929, and carried out geological work in northwestern QC. In 1929, was appointed professor of geology and department chairman at the University of Saskatchewan. Appointed dean of engineering in 1961. On retirement in 1963, remained as dean emeritus, and continued as director of the Institute for Northern Studies. Author of numerous reports, papers, and memoirs on geology and related topics. **CIM:** Member 1922; Councillor, District 4, 1945-47; Vice President 1954-55; Chairman of Geology Division 1948-49; President 1961-62; awarded the Barlow Memorial Medal in 1952. Member of Engineering Institute of Canada, Association of Professional Engineers of Saskatchewan, Association of Petroleum Geologists. Fellow of the Royal Society of Canada. Encouraged interdisciplinary work on Canada's northland.

John Ross Bradfield, 1962-1963

B. April 5, 1899, Morrisburg, ON. d. October 29, 1983, Toronto, ON. Graduated in civil engineering at McGill University in 1922. Honourary degrees from McGill University, Université de Montréal, and Université Laval. Spent his entire career at Noranda Mines except for a short time in the US, and with A.E. Wheeler, when in 1926, he began working on the design of the Noranda mine plant and smelter. Joined Noranda Mines in 1930 as plant engineer, became its secretary in 1938. He became successively, president in 1956, chairman and chief executive officer in 1962. In 1968, relinquished the latter post and remained as chairman. Retired in 1974, remaining honourary chairman and honourary director. **CIM:** President 1962-63; Life Member 1972; awarded the Inco Medal in 1973. President of the Mining Association of Canada, and Founding Director of the Canadian Mineral Industry Education Foundation. Companion of the Order of Canada in 1973. Inducted into the Canadian Mining Hall of Fame in 1990. Member of the Order of Canada. Took a great interest in the training and advancement of young engineers, and was personally responsible for the creation of a mining faculty at Université Laval.

William James Johnson, 1963-1964

B. Slocan, BC. d. circa 1978, Florida. Graduated from McGill University in 1923 with a BSc in civil engineering. Joined John S. Metcalf Co., Montreal, working on projects in Halifax, St. John, Port McNicholl, Houston, Texas, and Buenos Aires, Argentina. In 1934, joined Johnson's Co., Thetford Mines, QC, as mechanical superintendent of open-pit operations. In 1937, became mine superintendent, then manager in 1946, and managing director in 1954-1966. **CIM:** Member 1934; Chairman of Thetford Mines Branch; Councillor for 14 years; Chairman of Executive Committee for 5 years; President 1963-64; Life Member 1971.

Ralph D. Parker, 1964-1965

B. 1898, Lockeford, CA. d. July 31, 1983, Toronto, ON. Graduated from University of California with a BSc in mining engineering in 1921. Awarded LLD by University of California and by Laurentian University. First worked in placer operations in California and BC. In 1921, joined McIntyre Porcupine Mines Ltd. in Timmins, ON, as explorations engineer, and then became assistant mine superintendent. In 1928, was superintendent of Inco's Creighton mine, and later, the Frood mine. Became Inco's superintendent of mines in 1931, general manager of Canadian operations in 1954, vice president in 1955, and senior vice president in 1960-63. Remained as a consultant after retirement in 1963. Director of several companies. Oversaw the discovery and development of the Thompson, MB, nickel deposit and the transition from open-pit to underground mining in Sudbury with the introduction of bulk mining methods and equipment. Promoted the design and development of innovative mining method equipment. **CIM:** Member 1932; Chairman of the Metal Mining Division 1947-48; President 1964-65; Life Member 1966; Member of Fifty-Year Club; awarded Inco Medal in 1932. Member of AIME, receiving its Saunders Gold Medal in 1967. Member of several professional societies. Director of the Ontario Mining Association, the Canadian Metal Mining Association, and the Canadian Council of the International Chamber of Commerce. Inducted into the Canadian Mining Hall of Fame in 1991. Awarded the Inco Chairman's Citation for achievements as an engineer, administrator, and executive.

Frank A. Forward, 1965-1966

B. 1902, Ottawa, ON. d. August 6, 1972, Vancouver, BC. Graduated from University of Toronto with a BASc in chemical engineering in 1924. In 1924-29, was a smelter operator and research metallurgist with Consolidated Mining and Smelting Co. Ltd. (Cominco) at Trail, BC. Moved to Australia in 1929 where he was assistant smelter superintendent at Mount Isa Mines Ltd. in Queensland until 1934. In 1935, returned to Canada as a metallurgist with B.C. Nickel Mines Ltd. Joined the University of British Columbia as assistant professor of metallurgy in 1935, becoming professor in 1941, and department head in 1945-1964. Director of the Canadian Uranium Research Foundation in 1960. First director and organizer of the Science Secretariat in the Privy Council Office. Retired from the Science Secretariat and from the University in 1967. Consultant for Sherritt Gordon Mines Ltd. Inventor of the ammonia pressure leaching process used for the extraction of nickel and cobalt by Sherritt Gordon. Co-inventor of the uranium leaching process used at the Eldorado mine in Beaverlodge, SK. Canada's foremost hydrometallurgist. Committed to education and to the promotion of careers in mining and metallurgy. Often flew his own airplane. **CIM:** Junior Member 1924; Member 1934; Councillor 1939-41; Vice President 1943-45; Chairman of the Metallurgy Division 1950-51; President 1965-66; received the Alcan Award in 1969, and Inco Medal in 1955. President of BC Association of Professional Engineers in 1948, Dominion Council of Professional Engineers in 1949. Recepient of many other honours and awards including the McCharles Prize, EIC Leonard Medal, IMM Gold Medal, and James Douglas Gold Medal of AIME. The Frank A.

Forward Building at the University of British Columbia was named in his honour. An inspired scientific, professional and academic leader.

George Wheeler Govier, 1966-1967

B. June 15, 1917, Nanton, AB. Graduated from the University of British Columbia with a BASc in chemical engineering in 1939, an MSc from the University of Alberta, and a DSc from the University of Michigan. From 1940 to 1963, was successively instructor, then professor of chemical engineering, head of the Department of Chemical and Petroleum Engineering, and finally dean of the Faculty of Engineering at the University of Alberta. In 1948, appointed to the Petroleum and Natural Gas Conservation Board, later renamed the Oil and Gas Conservation Board, and in 1962, became its chairman. In 1963, was appointed to the Faculty of Engineering at the University of Calgary. **CIM:** Member 1950; Chairman of the Petroleum and Natural Gas Division 1950-51; Chairman of the Calgary Branch 1964-65; President 1966-67; awarded the Selwyn G. Blaylock Medal in 1971. President of Association of Professional Engineers of Alberta 1958-59. Member of the Engineering Institute of Canada, Chemical Institute of Canada (CIC), and the American Institute of Chemical Engineers. Recipient of the R.S. Jane Memorial Award (CSChE), and the Award of Merit of the Canadian Natural Gas Processing Association. Continues to be active in the petroleum and natural gas industry as a consultant.

Edward Dow Brown, 1967-1968

B. March 7, 1907, Nova Scotia. d. June 27, 1992, Halifax, NS. Attended Nova Scotia Technical College. Graduated from Dalhousie University in 1929. Awarded an Honourary DSc from Nova Scotia Technical College. Worked in the hydroelectric industry before becoming a mining engineer with Adanac Gold Mines in Rouyn, QC. In 1933, joined Noranda Mines Ltd., working on surface and underground. Moved to Bankfield Consolidated Mines Ltd., Geraldton, ON, in 1934, becoming assistant manager. In 1941, moved to Steep Rock Iron Mines at Atikokan, ON, as superintendent of mechanical and electrical work. In 1943, returned to Nova Scotia as chief engineer for Barytes Mines Ltd. at Walton. Consultant for Mosher Limestone Ltd. during quarry development and mill construction. In 1946, joined National Gypsum (Canada) Ltd., becoming vice president, then general manager. **CIM:** Member 1950; MSNS Councillor, MSNS President 1955-56;

CIM President 1967-68, during Canada's Centennial year; Life Member 1981. Member and council member of Engineering Institute of Canada, president of Professional Engineers of Nova Scotia, member of Board of Governors of Nova Scotia Technical College.

Paul-Emile Auger, 1968-1969

B. September 30, 1908, Ste. Croix, Lotbinière, QC. Graduated from Université Laval with a BA in 1929, and a BSc in chemistry in 1933. A researcher and instructor at the Superior School of Chemistry. Graduated from Queen's University in 1936 with a BSc in geology, and earned a DSc at the Massachusetts Institute of Technology. Spent his summers working on survey and geological crews throughout Quebec and Labrador. Progressed from geologist with the Quebec Department of Mines in 1938-41, to mine geologist at the Mine School in Val d'Or, QC, in 1939-42, and then a lecturer, assistant professor, and professor of economic geology at Université Laval in 1941-60. In 1960, was appointed deputy minister of the Department of Mines of Quebec, and in 1961, deputy minister of the Department of Natural Resources. Secretary general of ACFAS in 1946-52. Wing Commander in the RCAF, and Commanding Officer of Laval University Squadron, 1948-59. Has published extensively on geology. **CIM:** Member 1941; Geology Division Chairman 1960-61; Chairman of Quebec City Branch 1954-55; Vice President, District 2, 1965-66; President 1968-69. Awarded the Leonard Medal of the Engineering Institute of Canada in 1942. Member of several professional associations. Fellow of the Royal Society of Canada.

James Merritt Harrison, 1969-1970

B. September 20, 1915, Regina, SK. d. July 6, 1990, Ottawa, ON. Earned a BSc in geology from the University of Manitoba in 1935. Obtained an MA at Queen's University in 1941, then a PhD in 1943. Awarded Honourary DSc by six Canadian Universities. Joined Geological Survey of Canada in 1943 as a field geologist. Appointed chief of Precambrian Division in 1955, and director in 1956. Assistant deputy minister (research) for Department of Mines and Technical Surveys in 1964. Later his title changed to assistant deputy minister (mines and geosciences) in the new Department of Energy, Mines, and Resources. Was senior assistant deputy minister in 1972. **CIM:** Junior Member 1944; Member 1949; General Chairman of

AGM 1952; Vice President, District 3, 1959-60; President 1969-70; Life Member in 1980; awarded Selwyn G. Blaylock Medal in 1966. Fellow of the Royal Society of Canada. President of the Geological Association of Canada in 1960-61, receiving its Logan Medal in 1969. Regional vice president of the Society of Economic Geologists in 1962. First president of the International Union of Geological Sciences in 1961-64. Outstanding Achievement Award of the Public Service of Canada in 1970. Companion of the Order of Canada in 1971. Chaired the Canadian Commission for UNESCO in 1985-88. Chemist, geologist, teacher, and in later years, concerned with science organization, administration, and policy. Cultivated a deep understanding of Canada's northland and its peoples. One of the "best and brightest" of his generation. The mineral harrisonite is named after him.

Harold M. Wright, 1970-1971

B. December 10, 1908, Winnipeg, MB. d. December 11, 1997, Vancouver, BC. Graduated with a BSc in geological engineering from the University of Utah, and MA in geology and an MSc in metallurgical engineering from the University of British Columbia. In 1931-22 worked for the Geological Survey of Canada, then as a field engineer for the Denver Equipment Company (Canada) Ltd., and subsequently various mining companies and consultants in Canada and abroad. In 1946, established the internationally known Wright Engineers Ltd. He was founding president of Western Mines Ltd. (now Westmin Resources) which acquired various mining properties in British Columbia and brought them into operation. **CIM:** Member 1936; Chairman of Vancouver Branch, and the BC Section; Vice President, District 6, 1951-52; Chairman of Metallurgy Division 1952-53; President 1970-71; first recipient of the McParland Memorial Award in 1974. Member of the Association of Professional Engineers of BC, and the Association of Consulting Engineers of Canada. Inducted into the Canadian Mining Hall of Fame in 1993. Awarded the Gold Medal of the Institution of Mining and Metallurgy (UK). An outstanding athlete, he competed for Canada as a sprinter at the 1932 Olympic Games in Los Angeles. Member and President of the Canadian Olympic Association.

Thomas Herbert Patching, 1971-1972

B. October 2, 1915, Endon, AB. Graduated from the University of Alberta with a BSc in mining engineering. Worked as a mining engineer for

Inco, and for the Hudson Bay Mining and Smelting Co. In 1947, appointed assistant professor, then professor of mining engineering at the University of Alberta. Consultant on mining properties and mining methods. Conducted studies on coal outbursts, coal strength, and coal gas relationships. **CIM:** Member 1937; Chairman of Coal Division 1967-69; Chairman of the Northern Alberta Branch; Vice President, District 5, 1969-70; President 1971-72; served on several CIM committees; received the Coal Award in 1985. Member of several professional associations including the Alberta Association of Professional Engineers.

John H. Schloen, 1972-1973

B. February 25, 1909, Brooklyn, NY. d. August 6, 1997, Montreal, QC. Graduated in chemical engineering from McGill University in 1932. Began his career with Canadian Copper Refiners Ltd. (CCR Division of Noranda Inc.) in 1932, becoming plant superintendent in 1950, and president in 1968. Authority on electrolytic copper refining. **CIM:** Member 1951; Chairman of Montreal Branch 1956-57; Vice-Chairman, Executive Committee of Council, 1961-62; Chairman, Executive Committee, 1962-63, 1968-70; Councillor, District 2, 1961-63; Vice President, District 2, 1968-70; President 1972-73; served on the Finance Committee; Chairman of the Canadian Mining and Metallurgical Foundation, 1972-1991; Distinguished Lecturer, 1978. President of Laval Industrial Association. Member of Council, Montreal Board of Trade. Committee chairman, International Copper Research Association, and Selenium- Tellurium Development Association. Member of TMS-AIME, Institute of Metals, Electrochemical Society, the Institution of Mining and Metallurgy (UK).

James Parker Nowlan, 1973-1974

D. October 23, 1986, Halifax, NS. Graduated from Acadia University in 1928, and earned an MSc and PhD from University of Toronto in 1930 and 1935. Engaged in geological field work in Canada, Rhodesia, and Brazil from 1928-1939. Manager of Yann Gold Mines Ltd. in 1929-42. Supervisor at Canadian Industries Ltd., Brownsburg in 1942-45. Chief geologist of Cochenour Willans Mines Ltd. in 1945-49, then of Dominion Gulf Co. of Toronto in 1940-1953. Executive vice president of McPhar Geophysics Ltd. in 1953-58. Deputy minister of mines for Nova Scotia in 1958-73. **CIM:** MSNS President 1968-69; CIM President 1973-74. President of the Geological Association of Canada in 1964-65. Member of Board of Governors of the Canadian Institute for Radiation Safety in 1983-86.

William M. Gilchrist, 1974-1975

B. July 23, 1909, Weyburn, SK. Graduated from Queen's University in 1936 with a BSc in mining and metallurgy. As a student, worked for several mining companies in Manitoba, Ontario, and Quebec. In 1936, joined Preston East Dome Mines Ltd. In 1941, enlisted in the Royal Canadian Engineers and served with the 4th Canadian Armoured Division overseas, earning the rank of Staff Captain. After returning to Canada, worked for several mining companies, before joining Eldorado Mining and Refining Ltd. in 1952. In 1955, was appointed mine manager at Beaverlodge. Subsequently promoted to vice president of western operations. In 1958, became president of Eldorado Nuclear Ltd., and two of its subsidiaries, Eldorado Aviation and Northern Transportation. Currently president of W.M. Gilchrist and Associates Ltd. CIM: Member 1937; President 1974-75. Director of the Mining Association of Canada, Canadian Nuclear Association, and the Royal Canadian Geographical Society.

Patrick D. O'Connell, 1975-1976

B. January 16, 1927, England. Graduated from the University of Birmingham in 1950 with a BSc in chemistry. Served two years in the Royal Artillery. In 1950-54, worked in the oil and gas industry in Trinidad. From 1954 to 1957, was production and reservoir engineer with Iraq Petroleum in the Persian Gulf in Iraq, Qatar, Oman, and along the Trucial Coast. Involved in all phases of petroleum and natural gas production in western Canada since 1957. District production superintendent with Amoco Canada Petroleum Co. in 1976, then appointed manager of the new department of Public and Government Affairs. **CIM:** Councillor, District 5, 1969-71; Vice President 1971-73; Chairman of Calgary Petroleum Section; President of the Petroleum Society 1975-76; awarded the Petroleum Society's Distinguished Service Award. Recipient of the Queen's Silver Jubilee Medal. Member of SPE (AIME).

W. Mel Bartley, 1976-1977

B. 1910, Winnipeg, MB. d. August 12, 1986, Thunder Bay, ON. Resided in northwestern Ontario most of his life. Earned a BSc from St.

John's College, Winnipeg, an MSc from the University of Manitoba, and a PhD from the University of Toronto. As a student, worked for the Ontario Department of Mines. Chief geologist and production engineer for Steep Rock Iron Mines Ltd. in 1940-48. Formed his own consulting firm of M.W. Bartley and Associates in 1952. Exploration manager for Cliffs of Canada Ltd. in 1954-73. First principal of Lakehead Technical Institute, and founder, and first chairman of the Board of Governors, of Lakehead University. Lectured at Lakehead until 1985. Honourary professor in 1971, fellow of the University in 1981, and principal emeritus in 1985. **CIM:** Member 1935; Councillor, District 4, Vice President, District 4, 1955-56, 1969-72; President 1976-77; active in the Thunder Bay and Lakehead Branches. The present concept of District and Divisional technical meetings is due to his foresight. The Mel W. Bartley Award honours his efforts. Member of several professional societies but regarded the Institute as the most important organization in his career.

Ronald D. Lord, 1977-1978

B. December 26, 1910, Elginburg, ON. d. February 21, 1996, Toronto, ON. Graduated with a BSc in metallurgical engineering from Queen's University. Received an LLD from Queen's University in 1972. Joined the staff of Preston East Dome Mines, eventually became mill superintendent, manager and managing director. Joined Rio Algom in Toronto, later appointed managing director of a number of mines at Elliot Lake including Pronto and Milliken. Directed the construction and operation of the mines, and initiated new and improved techniques for mining, milling, and plant maintenance. Retired as vice president, research, of Rio Algom Ltd. in 1976. Appointed research director of Mining Research Organization of Canada (MIROC) in 1976. **CIM:** Member 1945; active in the Timmins and Toronto Branches; Vice President, District 3, 1973-75; President 1977-78; Life Member 1978; Fifty-Year Club 1995. President of the Ontario Mining Association.

Jean-Paul Drolet, 1978-1979

B. July 15, 1918, Quebec City, QC. Attended the Seminary of Quebec. Obtained a BA and a BSc in mining engineering from Université Laval. Earned an MSc in 1947 at Columbia University, NY. Honourary DSc from McGill University in 1978, and Laurentian University in 1979. Employed by Quebec Department of Mines before working in the exploration and development of iron ore in northern Quebec. Assistant to the president of Quebec Cartier Mining Co., served as director and executive officer. One of the founders of Port Cartier and Gagnon. In 1963, became assistant deputy minister (research) in the Department of Mines and Technical Surveys. After it reorganized, was appointed senior assistant deputy minister (international minerals) Department of Energy, Mines and Resources. Represented Canada at international conferences and meetings related to mining and minerals. Led mining missions to foreign countries, including the first Canadian Minerals and Metals Mission to the People's Republic of China (1972). **CIM:** Member 1944; Life Member 1982; Distinguished Lecturer in 1968; Vice President 1968-69; President 1978-79; Chairman of the North Shore and Ottawa Branches; served on many Institute committees; received the Past Presidents' Memorial Medal in 1964 and Distinguished Service Medal in 1984. Board member of Uranium Canada Ltd., Royal Canadian Mint, Centre for Resource Studies at Queen's University, Canadian Permanent Committee on Geographical Names, and Royal Canadian Geographic Society. Published extensively on mining technology and mineral economics. Guest lecturer at McGill University. He and his wife, Françoise Desrochers-Drolet, enamel artist and designer (with the French medallist Élie Pellegrin), conceived and endowed the Order of Merit of Sancta Barbara in 1967, to recognize women who have made a major contribution to the welfare and development of a Canadian mining community.

E.H. (Ted) Caldwell, 1979-1980

B. November 24, 1919, Killam, AB. Obtained a teacher's certificate from Camrose, AB, normal school, and taught for three years before joining the RCAF. In 1949, obtained a BSc in mining engineering from the University of Alberta. Joined the exploration group of Cominco's Mines Division at Trail, BC, and in 1955, appointed superintendent, western district. Became assistant manager, and in 1978, manager, exploration for Cominco Ltd. Retired in 1984. Enjoys curling, golf, tennis, and gardening. **CIM:** Member 1950; Chairman of Trail Branch 1959; Vancouver Branch member starting in 1963; Director and Chairman of BC Section 1963-65; Councillor, District 6, 1971; Vice President, District 6, 1972-74; President 1979-80. Vice president of B.C. and Yukon Chamber of Mines 1967-70. Member of Association of Professional Engineers of B.C.

Donald Earle G. Schmitt, 1980-1981

B. October 21, 1915, Kitchener, ON. Graduated from the University of Toronto with a BASc in mining engineering in 1937. Worked in various capacities at Sherritt Gordon, Bralorne, Kerr Addison, and Preston East Dome mines. Enlisted in 1943 in the Royal Canadian Engineers, becoming a Lieutenant. In 1950, joined the Noranda Group as mine superintendent at Pamour Porcupine Mines. Appointed successively as manager of Pamour in 1953, assistant manager, Horne Mine in 1956, assistant general manager, Noranda Mines Ltd., Toronto, in 1962, general manager in 1965, and vice president, mines, in 1966. President of Orchan Mines Ltd. and Central Canada Potash Co. until they amalgamated with Noranda Mines. Was responsible for the operation of Noranda's mines in Canada and the United States. Chairman of the Board, Pamour Porcupine Mines. **CIM:** Member 1938; Chairman of Toronto Branch 1968; Vice President, District 3, 1975-77; Chairman of John T. Ryan Safety Committee in 1968-77; Life Member 1977; President 1980-81; awarded the Inco Medal in 1984. President of the Association of Professional Engineers of Ontario, president of Quebec Metal Mining Association, chairman of the Canadian Mineral Industry Education Foundation, director and vice president of the Gold Institute. Well-known to people in mining communities, he helped improve their quality of life by supporting recreational activities, and participating on administrative boards.

Noel A. Cleland, 1981-1982

B. December 21, 1929, Sydney, Australia. Graduated from the University of Sydney in mining engineering in 1950. Employed for three years in coal and oil industries. Moved to Canada in 1954. In 1954-56, worked for California Standard Oil Ltd. (now Chevron Standard Ltd.). Joined Hudson Bay Oil and Gas Ltd. in 1957, working in Calgary and Edmonton in production and reservoir engineering. Responsible for the application of remote supervisory techniques in the oil and gas industry, and developed computer programs for engineering. In 1966, joined Sproule Associates Ltd. Appointed vice president, engineering, in 1972, and was president from 1983-1992. Retired in 1994 as director. Currently an independent engineering consultant and corporate director. Author of numerous papers on automation, natural gas engineering, economics, and Canadian petroleum reserves. **CIM:** Member 1963; Chairman of the Petroleum Society 1976; Councillor 1978; Vice President, District 5, 1979- 80; President 1981-82; Distinguished Lecturer 1974; received the Petroleum Society Distinguished Service Award in 1984, and Robert Elver Mineral Economics Award in 1991. Member of several professional societies. President of the Association of Professional Engineers, Geologists and Geophysicists of Alberta.

R. Heath Gray, 1982-1983

B. April 18, 1914, Sydney, NS. d. April 10, 1994, Montreal, QC. Graduated from McGill University with a BSc in mining engineering in 1936, and a PhD in geology in 1940. After graduation, worked for several years in a variety of positions, from underground surveyor with Dosco in Sydney, and Canadian Collieries (Dunsmuir) Ltd. on Vancouver Island, to exploration work for Standard Oil of B.C. Ltd. In 1943, joined Alcan at Arvida, working in the magnesium pilot plant. In 1945, became a sales engineer, and was on loan to the Department of Defence Production before rejoining Alcan in 1952. Rose to vice president, chemical products, Alcan Smelter Services Ltd. in 1970, before retiring in 1975. Co-founded Gray McKinlay Associates with Donald McKinlay, consultants to the chemical, metallurgical, mining and related industries. **CIM:** Junior Member 1942; Member 1952; Chairman of Montreal Branch 1953; Chairman of 56th AGM; Vice President, District 2, 1972-74; served for many years on various CIM committees; President 1982-83. The son of Francis W. Gray, CIM President in 1931-32.

Peter V. Young, 1983-1984

B. December 21, 1930, Glasgow, Scotland. Graduated from Imperial College (London) as an Associate of the Royal School of Mines. Worked in various mines in England, Wales, Sweden, and Zambia. Came to Canada in 1956. Joined the Iron Ore Company of Canada (IOC), becoming a specialist in mining engineering and long-range planning. Implemented the company's first training program. Was chairman of the Amalgamated School Board and responsible for schools in northern communities where IOC was the main employer. In 1968, joined Memorial University and developed a Cooperative Education Program in the Faculty of Engineering. In 1970, was university director of coordination, and oversaw the Cooperative (Work) Programs and placements of graduates. Active in a wide variety of research projects. In 1979, launched Probe Publishers Inc., which produced a national careers newspaper for universities, followed by Probe Research Management in 1981. **CIM:** Member 1956; Councillor 1964-65; Chairman of Knob Lake Branch 1958-59, Labrador City Branch 1964-65, and Newfoundland Branch 1971-72; served on Education and Technical Program Committees; President 1983-84. A professional-level photographer, and sings with the St. John's choir.

Walter J. Riva, 1984-1985

B. May 9, 1922, Canmore, AB. Served with the RCAF during World War II. Graduated from the University of Alberta with a BSc in mining engineering in 1949. From 1947 to 1972, worked for Great West Coal Co., West Canadian Colleries, and Canmore Mines Ltd., becoming executive vice president and chief executive officer of the latter. In 1972, joined Denison Mines Ltd. as vice president. In 1973, was vice president of Kaiser Resources Ltd. in Sparwood, BC, becoming president of its coal division in 1979. When Kaiser Resources became B.C. Coal Ltd. in 1980, he was appointed president and chief executive officer, and then chairman in 1981. Retired in 1983. Director of several corporations including Investment Corp., and Canarc Resources Corp., Westar Mining Ltd. and B.C. Resources. Invented the patented form-coke process. Co-invented a pelletizing process. Author of numerous technical papers. **CIM:** Member 1952; Chairman of Coal Division 1963-65; Councillor 1965-67; Vice President, District 5, 1973; President 1984-85; Distinguished Lecturer 1970; awarded the Selwyn G. Blaylock Medal in 1973, and the Coal Award in 1991. A member of several professional societies. Chairman of the Coal Association of Canada. Fellow of the Institute of Mining Engineers (UK).

Steven W. Harapiak, 1985-1986

B. April 12, 1933, Cowan, MB. Graduated from the University of Manitoba in 1959 with a BSc in mechanical engineering. Joined Edorado Mining and Refining Ltd. as maintanence engineer in 1959. From 1960-66 worked for Steep Rock Iron Mines Ltd. becoming maintenance superintendent. Engineering and maintenance superintendent for Alwinsal Potash in 1966-68, and production manager for Inland Cement from 1968-72. Project manager, and then general manager of Hudson Bay Mining and Smelting Co., from 1972 to 1977. Became vice president, uranium operations, Denison Mines Ltd. in 1977-79. Senior vice president, Noranda Mining Inc., in Toronto, in 1979-81, and president and chief executive officer of Potash Corp. of Saskatchewan in 1981-83. Recently involved in mining operations in eastern Russia. **CIM:** Member 1959; active in Mechanical/Electrical Division, and in several Branches; President 1985-86; Distinguished Lecturer 1978.

Charles Bruce Ross, 1986-1987

B. July 15, 1920, London, ON. Served with the Royal Canadian Engineers in 1940-45, retired as Captain. Graduated with a BSc in mining engineering from Queen's University in 1947. From 1947 to 1961, was mine supervisor at Hollinger Mines in Timmins, later becoming vice president and general manager and executive vice president. Appointed president of Labrador Mining and Exploration Co. Ltd. in 1979, and vice president, minerals, Norcen Energy Resources Ltd. prior to retirement. **CIM:** Junior Member 1948; Member 1954; Life Member 1984; Chairman of Porcupine and Toronto Branches; Chairman of John T. Ryan Trophies Committee in 1978-1981; President 1986-87; General Chairman of the AGM in 1988; served on many CIM Committees including by-laws. President of Ontario Mining Association in 1976, and president of Association of Professional Engineers of Ontario in 1984-85. He is the nephew of James G. Ross who was CIM President in 1935-36.

Lindsay Milne,1987-1988

B. November 12, 1937, Cambridge Bay, Victoria Island, BC. Graduated from the University of Saskatchewan in 1959 with a BSc in geological engineering. Attended the Banff School of Advanced Management in 1970. From 1959 to 1983, employed by Amoco Canada Petroleum Co. Ltd. and Amoco Production Co. (US). Began in the Drayton Valley region of Alberta, and held positions in Calgary and in Tulsa, OK, successively becoming chief engineer, district superintendent, division production manager, production manager, vice president of production, and then regional production manager, Houston east region. In 1983, joined Canterra Energy Ltd. as senior vice president of exploration and production for western Canada. Currently with Canadian Delta Exploration Ltd. **CIM:** Member 1971; Chairman of the Petroleum Society 1977-78; Councillor, District 5, 1981-82; President 1987-88; served on several Committees; awarded Past Presidents' Memorial Medal in 1979. Member of the Canadian Petroleum Association, Society of Petroleum Engineers of AIME, the Arctic Institute of North America, and other professional groups.

René Dufour, 1988-1989

B. June 25, 1931, Chicoutimi, QC. Graduated with a BSc in 1954 from École Polytechnique. First worked as a field engineer in charge of exploration in Quebec. In 1958, joined Quebec Cartier Mining Co., successively becoming planning engineer, pit engineer, chief engineer of mines, and mine superintendent. In 1965, appointed professor of mineral engineering at École Polytechnique. Involved in planning, evaluation, and development of several mineral deposits in Canada and elsewhere. Took part in several missions for the World Bank and the Canadian International Development Agency. Consultant to private industry, and to governments in Africa and South America. Special consultant to the Quebec Securities Commission. Executive director of the Polytechnique-McGill Centre on Automation and Robotics in Mining, and coordinator of the Cooperative Polytechnique-McGill BSc degree in mining engineering. In 1998, president of Niocan Inc., and director of Noranda Inc. and other companies. **CIM:** Member 1955; Chairman of several Committees including Education, Membership, Finance, Foundation; Chairman of the Technical Program for the AGM and the 10th Commonwealth Mining and Metallurgical Congress in 1974, and first Open-Pit Operators Conference in 1977; General Chairman of 1986 AGM; Chairman of Montreal Branch, 1974-75, 1985-86; Vice President, District 2, 1976-78; President 1988-89; awarded the Past Presidents' Memorial Medal in 1970; Distinguished Lecturer 1980-81; awarded Distinguished Service Medal in 1995. Received the Queen Elizabeth II Jubilee Medal in 1977. Member of the Ordre des ingénieurs du Québec, and several other professional associations.

John J. Laffin, 1989-1990

B. May 11, 1924, New Waterford, NS. Graduated from St. Francis Xavier University in 1945, and the Technical University of Nova Scotia (TUNS) in electrical engineering in 1947. Awarded an Honourary Doctorate in 1989 from TUNS. Held senior positions with the Mining and Transportation Division of the Dominion Steel and Coal Corp. In 1968 became assistant to the president of Nova Scotia Power Corp. Served as Nova Scotia deputy minister of mines and energy in 1980-1991. Currently president of Resource Management Consultants. Commissioner of the Supreme Court of Nova Scotia. **CIM:** Member 1947; Honourary Life Member and President (1967) of MSNS; Chairman of Halifax Branch 1973; received the MSNS Medal in 1985; CIM President 1989-90. President of the Association of Professional Engineers of Nova Scotia (APENS) in 1981-82. Received the APENS Outstanding Service Award in 1987, and the Gold Medal for Engineering Achievements in 1995. Chairman of the Canadian Atlantic

Section of the Institute of Electrical and Electronic Engineers in 1976. Chairman of the Federal-Provincial Committee on Atlantic Coal in 1982-1991. Active in community and charitable organizations. His wife, Theresa Laffin, was awarded the Order of Sancta Barbara in 1997.

Peter Tarassoff, 1990-1991

B. April 11, 1934, Montreal, QC. Graduated from McGill University in 1956 with a BEng in metallurgical engineering. Obtained a ScD in metallurgy from the Massachusetts Institute of Technology in 1962. First worked for Quebec Iron and Titanium Corp. in Sorel, QC. In 1964, joined Noranda Research Centre in Pointe Claire, QC. Held a variety of technical and managerial positions, and was appointed director of research and development in 1982. Appointed vice president and chief scientist of Noranda Inc. in 1988. **CIM:** Junior Member 1957; Member 1971; Technical Program Chairman of Conference of Metallurgists in 1972; President of Metallugical Society 1978; Fellow of CIM in 1987; President 1990-91; received the H.T. Airey Award of the Metallurgical Society in 1988, and the Silver Medal in 1992; Life Member 1997. In 1985, was the first Canadian president of the Minerals, Metals and Materials Society (TMS) of AIME. Director of AIME in 1984-86. Co-recipient of TMS Extractive Metallurgy Award in 1973, TMS Extraction Metallurgy Lecturer in 1984, elected Fellow of TMS in 1991. Recipient of James Douglas Gold Medal of AIME in 1995. Chairman of the Canadian Research Management Association in 1988-89. Chairman of the Minister's National Advisory Council to CANMET in 1988-91. Fellow of the Canadian Academy of Engineering in 1989. Member of the Ordre des ingénieurs du Québec, Minerals, Metals and Materials Society of AIME, Mineralogical Association of Canada. The mineral petarasite is named after him.

W.E. (Bill) Stanley, 1991-1992

B. December 25, 1939, Quebec City, QC. Graduated from Haileybury School of Mines in 1961, and earned a BSc in mining engineering from Michigan Technical University in 1964. In 1964-1967, worked for International Minerals and Chemicals as an industrial engineer, a senior mining engineer, and an employee development superintendent. In 1968, was a mining consultant for Stark Mining, and then vice president for J.D. Stark and Associates in Toronto. In 1970, was management consultant for Urwick Currie. In 1978, was appointed partner, and later director, of Coopers and

Lybrand Consulting Group, in charge of mining practices in Vancouver. Currently a retired partner, and independent consultant to the industry. Director of Tech Corp., Miramar Mining Corp., Luscar Ltd., Westmin Resources Ltd., and Gibraltor Mines Ltd. **CIM:** Member 1959; Distinguished Lecturer 1981; Chairman of Vancouver Branch 1983; Vice Chairman of the 1986 AGM; Vice President, District 6, 1988; President 1991-92; served on several important CIM Committees; recipient of Distinguished Service Medal in 1994, and Past Presidents' Memorial Medal in 1985. Member of several professional associations. Was a regular columnist for *Northern Miner* and *Mining Newsletter*. Authored numerous articles on topics from socio-economics to management and lifestyle.

A.D. (Tony) Anton, 1992-1993

B. December 19, 1943, Egypt. Graduated with a BSc in petroleum engineering from University of Cairo, Egypt, in 1966. Senior petroleum engineer with BP Canada from 1967 to 1974. Manager, reservoir engineering with Suncor from 1974 to 1982. Vice president, engineering and production, with AGIP Canada Ltd. in 1982-86. Then joined Husky Oil as manager, production evaluator. Currently with Pyramid Resources Ltd. **CIM:** Member 1977; Councillor, District 5; Chairman of Petroleum Society 1987-88; Vice President, District 5, 1989-91; President 1992-93; served on several CIM Committees.

David S. Robertson, 1993-1994

B. March 3, 1924, Winnipeg, MB. Graduated from University of Manitoba with BSc in 1946. Obtained PhD in geology at Columbia University in 1949. In 1942-1950, worked summers for the Geological Survey of Canada. From 1949 to 1951, was assistant professor of mineralogy and petrology at the University of Virginia, Charlottesville. In 1951-53, was a senior research geologist with Inco at Copper Cliff. Chief geologist, Angola, for the E.J. Longyear Co. in 1953-54. Appointed president of GMX Corp. in 1955, and directed the exploration that led to Stanrock Uranium Mines Ltd. Became vice president of Stanrock, and later a director. In 1962-64, was a partner of Robertson and Merritt. In 1965, founded David S. Robertson and Associates Ltd., a consulting firm which merged with Currie, Coopers and Lybrand. Retired in 1987, and is currently an independent consultant in the mineral industry. Director of several companies. **CIM:** Member 1947, Chairman of the Toronto Branch; District Councillor; Distinguished Lecturer in 1979-80; Chairman of Advisory Committee on Public Relations; President 1993-94; received Barlow Memorial Medal in 1970, and Distinguished Service Medal in 1996. Member of several professional associations. Served on several national and international committees.

William (Bill) Almdal, 1994-1995

B. May 26, 1939, Winnipeg, MB. Graduated from University of Manitoba in 1963 with a BSc in civil engineering. Worked for the Iron Ore Co. of Canada in Schefferville, QC, in 1963-73. Senior mining engineer at Shell Canada Ltd. in 1973-75. From 1975 to 1980, was senior project manager at McBean and Partners Engineering Ltd. Senior mining engineer at Petro Fina in 1980-82, and manager of engineering at Petro Canada in 1982-83. Senior project manager at Syncrude Canada Ltd., Fort McMurray, in 1983, becoming senior associate in 1991. Since 1995, a senior associate at Ziff Energy Group in Calgary. **CIM:** Member 1964; active in Schefferville, Sept Iles, Calgary, and Oil Sands Branches holding executive positions from 1965 to 1988; Councillor, District 5, 1982-84; Vice President, District 5, 1987-89; Chairman of several committees and technical meetings; President 1994-95; received the District 5 Distinguished Service Award in 1989, and the Distinguished Service Medal in 1990. Member of Society of Petroleum Engineers (AIME) and several other professional groups. Regional director of the Progressive Conservative Party of Alberta, 1990-95. Chairman, Fort McMurray Oil Sands Interpretative Centre Advisory Committe in 1989-1995. Helped crystallize the Whitehorse Mining Initiative.

Georgio Massobrio, 1995-1996

B. November 15, 1927, Italy. Graduated from the University of Genoa in electrical engineering. Studied advanced management at Harvard Business School. In 1951, began working for Italsider, became superintendent in 1959. Held positions as assistant general superintendent of operations for Italsider's Taranto Plant in 1961, and general superintendent of the Cornigliano Plant in 1965, was executive vice president in 1971. Moved to Canada in May 1977. Became vice president of Canadian Met-Chem in 1977-79. From 1980 to 1992, was president of Quebec Cartier Mining Co.; chairman of the board in 1992-94, honourary chairman from 1994 to 1996. **CIM:** Member 1978; member of the Montreal and Quebec North-East Branches for many years; President 1995-96; awarded the Inco Medal in 1989; general chairman of the

Organizing Committee of the Centennial AGM, and vice chairman of the CIM Centennial Corporation. Member of AIME, American Iron and Steel Institute, American Management Association. Director of the Mining Association of Canada, and Italian Metallurgical Association. Authored several papers on open hearth and basic steel making, and on the use of computers for production control and management information systems.

Donald Joseph Worth, 1996-1997

B. July 24, 1932, Haileybury, ON. Graduated with a BSc in mining engineering in 1954, and an MSc in 1955 from the University of Toronto. Worked for Broulan Reef Mines Ltd. as a labourer, miner, and sampler during the summers of 1949-53. In 1954, was a field engineer for Talvey Metals Ltd. In 1955-57, was mine foreman for Compania Minera Asarco, South America. Returning to Canada, was a hydraulics engineer for Lake Asbestos of Quebec Ltd. in 1958. Worked for Canadian Gypsum Company Ltd. as a mine engineer, and returned to Lake Asbestos of Quebec as assistant mine superintendent in 1960-64. Joined Canadian Imperial Bank of Commerce (CIBC) as a mining specialist in 1964, and was vice president from 1984 to 1997. Retired as corporate director in 1997. **CIM:** Student Member at University of Toronto; Member 1964; member of several Branches; served on many Committees; President 1996-97; received Robert Elver Mineral Economics Award in 1986, and Distinguished Service Medal in 1989. Recipient of the Distinguished Service Award of the Prospectors and Developers Association of Canada in 1991. Director of the Canadian Mining Hall of Fame. Enjoys curling, gardening, golf, and collects antique mining lamps.

Alexander (Sandy) M. Laird, 1997-1998

B. February 19, 1934, Invermere, BC. Graduated from the University of British Columbia with a BASc in mining engineering in 1957. Worked in various positions with mining and exploration companies in British Columbia, 1957-60. Joined Placer Dome Group in 1960. Held many positions from mine superintendent for Craigmont Mine (1960-64), and Endako Mine (1964-68), to mine manager, Marcopper Mining Corp., Philippines (1968-71), assistant to the vice president operations for Placer Development, Vancouver (1971-77), mine manager McDermitt Mine, NV (1974-77), manager of mines for Placer Amex, United States (1977-81), and general manager of Kidson Gold Mine, Australia (1981-85). Senior vice president, Placer Dome Inc. in 1995-97, and is currently executive vice president. **CIM:** Member 1958; Distinguished Lecturer 1990; President 1997-98; Member of several professional associations including SME (AIME), the Philippine Society of Mining, Metallurgical and Geological Engineers, and the Association of Professional Engineers and Geoscientists of British Columbia.

Yves Harvey, 1998-1999

B. January 2, 1950, Quebec, QC. Graduated with a BASc and an MASc in geological engineering from École Polytechnique in 1975. Obtained a PhD in economic geology from Université Laval in 1984. Joined the Société québécoise d'exploration minière (SOQUEM) in 1978 as a senior geologist. From 1987-89, worked for Roche Ltée, Groupe-Conseil de Sainte-Foy, QC, as director of the Division of Mines and Geology. Returned to SOQUEM as executive vice president, human resources. Currently, president and executive director. Director of SOQUEM, Novicourt, and Cambior. **CIM:** Member 1982; Chairman of Quebec Branch 1991-92; Councillor 1994-96; Vice President, District 2, 1996-97; member of the Strategic Planning Committee 1993-94 and 1996-97; Chairman of Committee on Relations between Districts and Societies/Divisions; awarded the Past Presidents' Memorial Medal in 1994, Medal of the Quebec Branch in 1995, and District 2 Distinguished Service Medal in 1996; President 1998-99. Member of the Ordre des ingénieurs du Québec, Association of Quebec Prospectors, and Prospectors and Developers Association of Canada.

APPENDIX II

INSTITUTE SECRETARIES

Secretaries

Benjamin Taylor A. Bell, 1898-1904

B. July 2, 1861, Edinburgh, Scotland. d. March 1, 1904, Ottawa, ON. Educated at Stewart's College in Edinburgh. Worked briefly for an engineering firm before arriving in Halifax, NS, in 1882. In 1883, was a clerk in the freight department of the Canadian Pacific Railway in Winnipeg. In 1884, joined the Manitoba Contingent of the Canadian Voyagers, bound for the Sudan. Earned the Egyptian Medal and Star. Returned to Montreal, QC, in 1885. Became editor of the *Athletic News*. Also worked for the Montreal *Gazette* and Montreal *Witness*. Editor of the *Canadian Mining Review* in 1886, sole owner by 1896. An enthusiastic sportsman and athlete, excelled in cricket. **CIM:** Founded and served as Secretary of the General Mining Association of the Province of Quebec, in 1891. Assisted the amalgamation of the Gold Miners Association and the Mining Society of Nova Scotia in 1892. Founded the Mining Institute of Ontario in 1894, and the British Columbia Association of Mining Engineers in 1895. Spearheaded the formation of the Federated Canadian Mining Institute in 1896, which became the Canadian Mining Institute in 1898. Secretary and chief organizer until his untimely death in 1904. The B.T.A. Medal was named in his honour. Inducted into the Canadian Mining Hall of Fame in 1998.

Harold Mortimer-Lamb, 1905-1919

B. May 21, 1872, Leatherhead, Surrey, England. d. October 30, 1970, North Burnaby, BC. Educated in private schools in England and Belgium. Came to Canada in 1889. First employed on a farm in Agazzi, BC. In 1891, was editor of the *Chilliwack Progress*. In 1894, went to the Boundary District, gained initial experience in prospecting. Became editor of the *Midway Advance*, then shortly after, established the *Boundary Creek Times* at Greenwood. A correspondent for *The Province*, a weekly periodical in Victoria. In 1897, was editor of the *BC Mining Record*, then managing editor and director. A correspondent for the *Engineering and Mining Journal*, and other technical publications. Undertook special publicity work for the British Columbia Department of Mines, helped organize a provincial mining association in that province, serving as secretary and director. (Not the present day BC Mining Association.) Editor of the *Canadian Mining Journal*. Editor of the *BC Miner* in 1928 (later the *Western Miner*). **CIM:** Member 1901; Secretary in 1905; Secretary-Treasurer in 1911; CIM official representative to the Mining Congress at Dusseldorf, Germany, in 1909. Edited the *Bulletin* from 1905 to 1919, and the *Transactions* from 1905 to 1921. In 1916, was Chairman of the Central Directing Committee of Technical Organizations cooperating with the Advisory Council for Scientific and Industrial Research (NRC). In 1919, resigned as CIM Secretary-Treasurer due to ill health, and returned to British Columbia as Secretary-Emeritus. In 1920, was appointed Secretary-Treasurer of the BC Division. Retired from Institute activities in 1945. Had a life-long interest in art. Wrote many articles on Canadian art for the *London Studio* and other periodicals. Charter member of the Montreal Art Club, and Vancouver Art Gallery. Member of the Canadian Federation of Artists. A director and vice president of the BC Art League. Honourary member of the London Salon of Photography. In 1945, he was the only Canadian Honourary Fellow of the Royal Photographic Society.

George Cleghorn Mackenzie, 1920-31

B. 1877, Kincardine, Bruce County, ON. d. August 22, 1931, Westmount, QC. Educated at Trinity College, Toronto, in 1896. Graduated from Kingston School of Mines in 1903 with BSc in mining engineering. Did field work in geology and mining under Alfred E. Barlow in central Ontario. As a student, worked in gold mines in BC. Became interested in metallurgy and worked for Canada Iron Furnace Co., Midland, ON. After graduation, was assistant blast-furnace superintendent at the Londonderry Iron and Mining Co. in Nova Scotia. Later promoted to superintendent. In 1907, was on staff of the Ontario Bureau of Mines as mining claims inspector in the Montreal River District. On the staff of the Mines Branch, Department of Mines, Ottawa until 1915. Appointed secretary to the Munitions Resources Commission of Canada in connection with wartime minerals and mineral products. In 1919, he became general manager of the Electro Steel and Engineering Co., Welland, ON. On January 1, 1921, assumed new duties as Secretary-Treasurer of the Institute. **CIM:** Member 1902; Councillor 1916-17. Devoted special attention to the *Bulletin*, improved its contents and appearance. In 1927, was appointed general secretary of the 2nd Empire Mining and Metallurgical Congress, responsible for arranging technical sessions and tours across Canada, and publishing five volumes of *Proceedings*. "A man of strong and pleasing personality, rarely endowed with a keen sense of humour and much tact."

E. J. Carlyle, 1931-49

B. October 15, 1877, Woodstock, ON. d. April 27, 1952, Victoria, BC. Graduated from McGill University with a BSc in mining engineering in 1904. Worked in metallurgy at Anaconda, MT, and later in Arizona. In 1907, spent four years in Russia in the Urals with the Kyshtim Corp. In 1912, made a brief inspection of smelters in Rio Tinto, Spain, then spent a year in Argentina. From 1914 to 1917, was metallurgical engineer in Russia for the Sissert Co., an Anglo-Russian corporation engaged in copper development. Due to political turmoil, returned to Canada, and in 1918 was smelter superintendent for the British America Nickel Co. at Sudbury, ON. From 1926 to 1931, was superintendent of the Shorey plant of the American Smelting and Refining Co. in Peru. Because of political and labour unrest, he returned to Canada. **CIM:** appointed Secretary-Treasurer in 1931; Executive Director in 1947; retired in 1949 as Secretary-Emeritus; Honourary Member of MSNS 1949; CIM Life Member 1951. Known as "Curly", he provided the Institute with "unremitting, conscientious, and devoted service. . . over almost two decades of troublous years." Responsible for a five-fold increase in student membership. A younger brother to CIM Founder, William A. Carlyle.

Carlyle (Jerry) Gerow, 1949-63

B. July 14, 1896, Bloomfield, ON. d. July 27, 1974, Bathhurst, NB. First worked as a public school teacher in Ontario and in Saskatchewan. During World War I, was in active service in France with the Canadian Field Artillery, and in England with the Royal Air Force. Graduated from Queen's University in 1922 with a BSc in chemical and metallurgical engineering. Worked for the Coal Sales Department of the Dominion Steel and Coal Corp. in Montreal, became district manager for Ontario. Specialist in steam-plant engineering and combustion. In World War II, supervised the bunkering of coal-burning convoy ships. Then was on loan to the Department of National Defence. Directed the supply of heating fuel, coal, oil, gas, and wood for all RCAF and RAF training schools in Canada. Awarded a Member of the British Empire (MBE). In 1946, was appointed manager-secretary of the Canadian Coal Operator's Association (1944) in Ottawa. **CIM:** Appointed Secretary-Treasurer July 1, 1949; retired in 1964 as Executive Director Emeritus. He guided a rapidly growing Institute through complex issues surrounding membership, conferences, technical groups, publications, and finances.

Richard Ethelred (Dick) Barrett, 1963-1965.

B. June 20, 1905, Cobden, ON. Graduated with a BSc in mining engineering from McGill University in 1926. Worked for three years for Cerro de Pasco Copper Corp. in Peru. Returned to Canada in 1929. Worked for the Britannia Mining and Smelting Co. Then entered the manufacturing industry at the Canadian Office and School Furniture Co., Preston ON. In 1933, was manager of Parkhill Gold Mines in the Michipicoten District, then from 1933 to 1947, manager of Central Patricia Gold Mines. In 1947, appointed professor of mining engineering at the University of Toronto, then head of the Engineering Department in 1948. In 1953, joined Eldorado Mining and Refining Ltd. as manager of the Beaverlodge Operation. Moved to Ottawa in 1955 to become director of ore production, until his appointment as Executive Director of CIM. **CIM:** Member 1934;.founding member of the Pickle Lake Branch, 1945; Chairman of Metal Mining Division 1949-50; Vice Chairman of Ottawa Branch 1956-57, and Chairman 1957-58; Councillor, District 3, 1961-63; appointed Executive Director 1963; Life Member 1971. Retired October 1, 1965. Member of AIME, and Ontario Association of Professional Engineers. Director of the Ontario Mining Association for a number of years. Known for "his wonderful enthusiasm and good humour."

Gordon Tapp, 1965-79

B. October 10, 1915, Port Arthur, ON. Graduated from Queen's University with a BSc in mining engineering in 1939. Worked several summers as a student at Larder Lake. Joined the engineering staff of Consolidated Mining and Smelting Co. (Cominco) at the Sullivan Mine in Kimberley, BC, in 1941. Worked as an underground surveyor. In 1942, was a testing engineer in the Testing and Research Department; in 1946, became assistant supervisor. In 1948, appointed cost engineer for the company's mine, concentrator, and fertilizer plant. **CIM:** Member 1939; founding member and first Secretary, then first Chairman, Sullivan Branch; appointed Assistant-Secretary of CIM in 1954, Secretary-Treasurer in 1963, and Executive Director in 1965; retired July 27, 1979 as Executive Director Emeritus. Along with numerous other tasks, was responsible for promoting membership, organizing the Annual General Meetings, and assisting the Branches in their administration. Introduced the concept of a Foundation to members of Council. Promoted the idea of District Meetings to Institute Vice Presidents as a way to better serve Branch members. A compassionate man whose understanding and tact were often displayed in the letters he wrote to members and their families.

Gordon F. Skilling, 1979-82

B. 1921 in Danville, QC. Graduated from McGill University with BSc honours geology in 1951. After four years with mining and exploration companies, worked ten years for commercial oil and gas and mining publications. **CIM:** Member 1949; appointed Secretary-Treasurer of CIM in 1965. Managed publications and introduced the *CIM Directory* in 1967, the *CIM Reporter* in 1975, and the District 3 Directory in 1980. Expanded services to CIM members and the Canadian mineral industry. Trained Institute staff in state-of-the-art magazine production. Served briefly as Secretary of the Foundation. Retire prematurely due to ill health and, with his wife Shirley, has enjoyed gardening, sports and travelling over the past 15 years.

Pierre Michaud, 1982-92

B. September 4, 1925, New Brunswick. d. April 26, 1996, Dorval, QC. Served in the Royal Canadian Navy during the latter part of World War II. Graduated with a BASc in mining engineering from Université Laval in 1950. From 1950 to 1957, was employed in several supervisory positions at East Sullivan Mines Ltd., a copper producer in Val d'Or, QC. In 1982, joined Atlas Copco Inc., starting as a sales engineer covering the Chibougamau mining camp and the Gaspé area. In 1959, transferred with the company to Sudbury as branch manager. In 1963, was sales manager for the Mining and Construction Division based in Montreal. Eventually became manager of marketing services at the company's head office in Dorval. **CIM:** Member 1965; Chairman of Montreal Branch, 1979-80; Councillor, District 2, 1981-82; appointed Executive Director, October 1, 1982; retired September 20, 1992. These years were turbulent ones for both the Institute and the mining industry, and Michaud will long be remembered for his strong leadership.

Yvan Jacques, 1992-

B. January, 30, 1937, Sherbrooke, QC. Graduated from Université de Montréal with a BASc in agricultural sciences in 1961, and from Université de Sherbrooke with an MBA in 1968. Spent several years in agricultural research, and in food-related industries in Montreal and area. From 1974 to 1978, was assistant deputy minister, marketing and economics, Quebec Ministry of Agriculture and Food, and from 1978 to 1979, was assistant deputy minister, finance and administration, for the Quebec Ministry of Natural Resources. In 1979-80, was vice president and general manager of Myriad Detergent and Chemicals, a wholly-owned subsidiary of Lever Brothers Ltd. From 1980 to 1986, was assistant deputy minister, Marketing and Economics Branch, Agriculture Canada, and from 1986 to 1988, headed its International Programs Branch. In 1988-92 was executive vice president of the Agricultural Institute of Canada (AIC). **CIM:** Appointed Executive Director September 1, 1992. A member of several professional associations, and director of several academic institutions, private companies, and corporations. Participated in several International Joint Economic Committees, and led technical-trade missions to several countries. Commander of the Order of Agronomic Merit from l'Ordre des agronomes du Quebec. An all-round athlete, and the first CIM Executive Director appointed from outside of the mining industry.